Collins *gem*

Carbon Counter

Mark

HarperCollins Publishers
77-85 Fulham Palace Road, London W6 8JB

www.collins.co.uk

First published 2007

Reprint 10 9 8 7 6 5 4 3

ISBN 978-0-00-724812-4

Editorial by Grapevine Publishing Services, London
Text by Mark Lynas
Design and graphics by Judith Ash

Printed in Italy by Amadeus S.r.l. on Cyclus Offset,
a 100% recycled paper

CONTENTS

INTRODUCTION

You've heard of counting carbs. This book is about counting carbon. It's not the health of your body that is the object of our interest this time, but the health of the planet. Carbon dioxide is the main gas responsible for global warming, and humans are producing 25 billion tonnes of the stuff every year, raising the temperature of the planet to dangerous levels. This book will help you bring down your personal contribution to this rather daunting problem.

According to the government's chief scientist Sir David King, global warming is the greatest problem facing humanity. It's therefore a bit of a tall order to solve it all on your own. But if enough people begin to count and reduce their carbon, the rate of climate change can be slowed down dramatically.

This could save countless species from extinction, as animals like the polar bear and Bengal tiger will have nowhere to go as the world warms. It could also save the lives of the millions of people who are at risk from rising seas in low-lying areas. But if we don't act, vast swathes of the globe could be rendered uninhabitable because of flood, drought and searing heatwaves.

So by ceasing to be part of the problem and instead becoming part of the solution, you can join a growing movement to cut back humanity's carbon emissions before it is too late – and many scientists suggest that the atmosphere's 'tipping point' will come within the next decade, giving us no time to dither before making dramatic cutbacks. The lifestyle change that this implies doesn't necessarily mean misery and sacrifice: low-carbon living means that we should all end up fitter and healthier too. So maybe – the planet aside – counting carbs and counting carbon aren't so different after all.

I
Welcome to the greenhouse

THE GREENHOUSE EFFECT

Contrary to what you might have heard, the greenhouse effect is not all bad. Without the heat-trapping effect of certain gases, the average temperature of the planet would be a decidedly chilly -18°C. Nor is it a new discovery. The Irish physicist John Tyndall was the first to realize, back in 1859, that without greenhouse gases in the atmosphere, things would quickly turn unpleasantly cold. As he wrote (a little melodramatically), without these heat-trapping gases, 'the warmth of our fields and gardens would pour itself unrequited into space, and the sun would rise upon an island held fast in the iron grip of frost'. (Tyndall did well on the lecture circuit.)

Tyndall correctly identified that the two most common gases in the air, nitrogen and oxygen, are transparent to heat radiation. Instead, it is gases present in much smaller quantities – water vapour, carbon dioxide, methane and others – which have the heat-trapping effect. No one disputes this 'natural greenhouse effect', as it can be demonstrated easily enough in the laboratory. The arguments (for climate change is nothing if not controversial) come later.

Not all greenhouse gases are born equal. Carbon dioxide gets most attention because it is the most abundant, and lasts a long time in the atmosphere. Water vapour is also a greenhouse gas, but its emissions are more or less irrelevant because it condenses out as rain in a matter of hours or days. (The exception to this is water vapour emitted high in the atmosphere by aircraft, where it can have a longer-term warming effect.) The table below details the different gases and their ability to trap heat – their 'global warming potentials' – as well as their atmospheric lifetimes.

GREENHOUSE GASES			
Gas	Chemical formula	Global warming potential (over a century)	Atmo-spheric lifetime (years)
Carbon dioxide	CO_2	1	100–1000
Methane	CH_4	23	12
Nitrous oxide	N_2O	296	114
Chlorofluorocarbons	CFCs (various)	6000–14,000	45–1700
Hydrofluorocarbons	HFCs (various)	12–12,000	0.3–260
Sulphur hexafluoride	SF_6	22,000	3200

'Global warming potential' means the potential each gas has to trap heat over a given period of time, just as different ratings are applied to measure the heat-trapping strength of different duvets on a bed!

The greenhouse effect can also be observed on other planets in the solar system. Venus has a hotter surface temperature than Mercury, despite being much further away from the sun. Because the Venusian atmosphere is almost entirely composed of carbon dioxide, this gives it a phenomenally strong greenhouse effect – keeping surface temperatures at 460°C, hot enough to melt lead.

All of this would interest no one other than chemists and astronomers if the amounts of these gases in the earth's atmosphere were to stay fixed. From the Industrial Revolution, humans began exploiting fossil fuels – coal, then oil and gas – for energy. These fossil fuels burn to produce carbon dioxide, which then accumulates in the atmosphere over the decades and centuries.

In 1750 there were 278 parts per million (ppm) of carbon dioxide in the earth's atmosphere. Now that figure has risen to 380ppm, and it is continuing to rise at 1.5 to 2ppm per year. Methane levels have risen from 700 to 1745 parts per billion, while nitrous oxide has jumped from 270 to 314ppb. The other gases – CFCs, HFCs and SF_6 – are entirely products of the industrial age, so did not exist in 1750. But now that they do, many will be with us for a very long time, as the right-hand column of the table on page 9 shows.

Put these changes in a geological context, and it all begins to sound a bit scary. Carbon dioxide levels haven't been this high on earth for millions of years. The temperature of the planet rose and fell with the cycles of the ice ages, but during the whole time carbon dioxide levels were lower than they are now. But as the graph on page 12 shows, temperature and carbon dioxide shadow each other very closely over the long term, suggesting that our increase in CO_2 will indeed be followed by a temperature increase.

WHAT ARE FOSSIL FUELS?

Fossil fuels are often termed 'buried sunshine', because, in essence, they represent energy captured from the sun by photosynthesis in ancient plants. By using fossil fuels, humans are getting an effective

Tip: If you think about it, all the energy humans have ever used has come from the sun. Cavemen burned wood in fires for heat and cooking, but the trees it came from used the sun's energy.

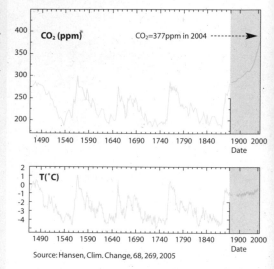

CO₂ AND ESTIMATED GLOBAL AIR TEMPERATURE

Source: Hansen, Clim. Change, 68, 269, 2005

Ice cores drilled by glaciologists from Antarctica provide a good record of temperature and greenhouse gas fluctuations over the ages. Two things are striking: first, how closely correlated temperature and CO_2 are; and second, how much higher levels are now than during the whole period of record.

energy subsidy from the past. Before the discovery of coal, oil and gas, humans – like other animals and plants – had to live only from the energy provided directly by the sun.

Coal: Coal is the fossilized remains of ancient forests. Many of these forests built up thick layers of peat underneath them, much as do forests today in hot tropical areas like Indonesia. Over millions of years, this peat became compressed by layers of sediment above it, and turned gradually to carbon-rich coal. A prime coal-forming era was the Carboniferous period (300–360 million years ago), which was named after the extensive coal beds found in Western Europe. Most of the electricity produced worldwide comes from coal.

Oil: Unlike coal, oil begins its formation in the sea. The dead remains of plankton accumulate in bottom sediments, where they are eventually buried at great depths and heated up by geothermal processes. This then 'cooks' the dead algae, releasing hydrocarbons, which companies such as Esso and BP later hope to drill out from reservoirs of oil trapped between impermeable rocks. If the oil is cooked at too high a temperature, it instead forms gas (see page 15). Saudi Arabia and Iraq have the largest proven oil reserves on the planet. Their oil was mostly formed during the Jurassic era, 200–145 million years ago. An incredible 80 million barrels of oil are consumed by humanity each day – that's nearly a thousand barrels per second.

SOURCES OF GREENHOUSE GASES

Carbon dioxide: mainly produced by the combustion of coal, oil and gas from fossil fuels. Also results from deforestation and land clearance, and from peat burning in the tropics.

Methane: produced in places where oxygen-free decomposition occurs, such as in rice paddies and landfill sites. Can also be released by leaky pipelines and burping cows.

Nitrous oxides: emitted from burning fossil fuels – in car exhausts, for example – and from agricultural fertilizers breaking down in the soil.

HFCs and other industrial gases: as it says, these are produced by industrial processes. HFCs were introduced as replacements for CFCs, which are now banned to protect the ozone layer. They are used in refrigerants, in aerosols and foam-blowing. SF_6 is used in high-voltage electrical switches, but also in that squishy bit of some training shoes, and inside tennis balls.

Gas: Much natural gas is overcooked oil, which is why oil and gas fields tend to be so closely associated. Some of the biggest gas fields lie in the Middle East and Russia. Gas is mostly transported via pipelines, though it can also be liquefied under high pressure and transported on ships. In its natural state, gas is odourless – small amounts of additives are included in the mixture in order to give it a characteristic smell, useful in detecting leaks. Leaked gas accumulating in buildings can cause deadly explosions.

GLOBAL WARMING TODAY

There is no longer any serious scientific doubt that global warming is already under way. The earth is now 0.7°C warmer than it was 150 years ago, before the Industrial Revolution began to change the amount of greenhouse gases in the atmosphere. The graph on page 16 shows how air temperatures have risen since 1850. You can see that the rise has not been uniform, but has come in two distinct blips – between 1910 and 1940, and between 1970 and today. This begs the question of why temperatures stabilized, and even cooled slightly, after the Second World War, when emissions of carbon were rising rapidly. Scientists think that the culprit may have been another pollutant, sulphur dioxide, which acts to counteract warming by reflecting some of the sun's heat. Thus 'global dimming' counteracted 'global warming', and the two forms of pollution to a certain extent cancelled each other out. Sulphur dioxide is now produced less, however, because it causes acid rain and has nasty effects on people's health.

GLOBAL AIR TEMPERATURE 2005 ANOMALY +0.48°C (MEAN 1961–1990)

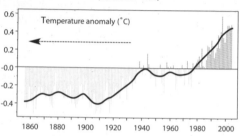

Source: © 2006, Climatic Research Unit, University of East Anglia. www.cru.uea.ac.uk

Tip: Sulphur dioxide was one of the main pollutants in the choking London 'pea souper' smogs of the 1950s, which were controlled by the Clean Air Act of 1956. It was produced as a by-product of burning fossil fuels (especially coal), because of small amounts of sulphur contaminating the fuel. Low-sulphur diesel is manufactured specially to reduce sulphur dioxide emissions.

FIVE HOTTEST YEARS IN DESCENDING ORDER

1:	2005
2:	1998
3:	2002
4:	2003
5:	2004

All five of the hottest years ever recorded were in the last decade.

Source: NASA (http://www.nasa.gov/vision/earth/environment/2005_warmest.html)

Global warming impacts so far

- Sea levels are rising at about 3mm per year, faster than during the whole of the last one hundred years. People in Pacific island nations like Tuvalu are already planning to leave their homelands, while cities like New York and London are threatened by rising tides.
- Mountain glaciers are retreating worldwide, from the Alps to the Andes. The Alps have lost half their glacier cover since 1850. In the Andes, the melting glaciers mean that essential freshwater supplies will soon be lost for ever.
- Snow cover in northern areas is decreasing, and permafrost in the Arctic is melting, causing damage to buildings and roads in Alaska and Siberia.

- The melting icecap on the Arctic Ocean is affecting animals that depend on it. Polar bears have been left without ice to rest on, and some have drowned while trying to swim long distances. Walruses have tried to climb into white fishing boats, mistaking them for the ice floes that have disappeared.
- Both Greenland and Antarctica are losing ice mass; Greenland now loses 250 cubic kilometres of water each year, adding to sea level rise.
- Major floating ice shelves have collapsed in Antarctica. The latest was Larsen B, which fell apart spectacularly in a few days in 2002, much to the shock of watching scientists.
- Droughts have increased around the world, as has the incidence of heavy rainfall, due to the speeding up of the water cycle with more energy in the atmosphere. In Mumbai, during the 2005 monsoon, nearly a metre of rain fell in a single 24-hour period (the highest total ever), killing 750 people with the resulting floods.
- Hurricanes and typhoons are getting stronger, with heavier rainfall and more destructive winds. Hurricane Katrina, which devastated New Orleans,

Tip: Hurricanes get their energy from the heat of the oceans, which evaporate their water into the swirling clouds that make up these terrifying storms. So as the oceans get hotter because of global warming, hurricanes have more 'rocket fuel' to boost their power.

was one of the strongest Atlantic storms ever observed. Half a million people have still not returned to the ruined city, and are considered 'climate refugees' by some scientists.

- Heatwaves have also increased in frequency and intensity. In 2003, severe heat killed up to 30,000 people across Europe. In Paris, morgues were overflowing with the corpses of elderly people, and some had to be stored in vegetable cooling warehouses.

GLOBAL WARMING OVER THE NEXT 100 YEARS

How severe global warming gets depends mostly on how far humanity lets greenhouse gases continue to spiral upwards – and that's where this book comes in. The Intergovernmental Panel on Climate Change, the main scientific body on this issue, states that temperatures will increase by 1.4ºC to 5.8ºC over the next hundred years. The key uncertainty here is not how the atmosphere will behave, but how humans will behave. Will seven or more billion humans continue to seek fossil fuel-hungry lifestyles, with cars, aeroplanes and all the rest? Or will they develop cleaner energy sources and tread more lightly on the planet? There is no clear answer to this yet – indeed the ultimate answer will depend on decisions made by each and every one of us, including you.

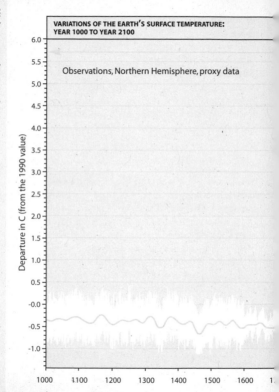

VARIATIONS OF THE EARTH'S SURFACE TEMPERATURE: YEAR 1000 TO YEAR 2100

Observations, Northern Hemisphere, proxy data

Departure in C (from the 1990 value)

6.0
5.5
5.0
4.5
4.0
3.5
3.0
2.5
2.0
1.5
1.0
0.5
-0.0
-0.5
-1.0

1000 1100 1200 1300 1400 1500 1600 1

Global instrumental observations

Projections

This graph shows the fluctuations in Northern Hemisphere temperature over the last one thousand years. You can see that changes have been relatively minor – the much-discussed 'Little Ice Age' and 'Medieval Warm Period' barely stand out. What does stand out is how unusual 20th-century warming has been in this long-term context. Even more striking are the projections for rapid warming over the next one hundred years.

1800   1900   2000   2100

Source: Intergovernmental Panel on Climate Change, Third Assessment Report, 2001. www.ipcc.ch

Here's what – if they happen – each of these degree rises in temperature might mean for the planet and our society:

One degree
- Deserts invade the High Plains of the United States, in a much worse repeat of the 1930s dustbowl. While the epicentre is Nebraska, states from Ontario in the north to Texas in the south suffer severe agricultural losses.
- Mount Kilimanjaro loses all its ice.
- The Gulf Stream switches off – perhaps plunging Britain and Europe into icy winter cold.
- Rare species, like forest-dwelling frogs, possums and tree kangaroos, are wiped out in the Queensland rainforest, Australia, as warming erodes their habitat.
- Coral reefs around the world suffer increasing losses from bleaching and are wiped out, with the Great Barrier Reef largely destroyed by 2030.
- Island nations submerge under the rising seas.

Two degrees
- Oceans turn increasingly acidic, further hitting coral reefs and endangering the marine ecosystem.
- One summer in two has heatwaves as strong as 2003's disaster in Europe.

THE GULF STREAM

The Gulf Stream is one part of a massive interlinked conveyor belt of currents that takes heat around the global oceans. Consider that London lies further north than Vancouver in Canada, while northern Scotland is at the same latitude as southern Alaska – but in both cases Europe has a much milder climate, thanks to the warm ocean currents we enjoy.

This current loses its heat in the North Atlantic and Arctic Oceans, and the water sinks to the bottom of the ocean bed. This happens because colder water is more dense and salty, and therefore heavier, driving the movement of the 'conveyor'. The fear is that global warming will lead to a fresher ocean surface in these sinking regions due to higher rainfall and melting ice, shutting off the crucial driving point of the ocean conveyor. Indeed, a slowing of the current has already been detected by scientists.

However, there is disagreement in the scientific community as to whether this slowing is just a blip or something more permanent. Most computer models do not suggest that the shutting down of the ocean circulation would lead to a new ice age in Europe, although our weather might still change dramatically. So the flash-freezing scenario played out in the movie *The Day After Tomorrow* is likely to remain fiction.

- Greenland tips into irreversible melt, accelerating sea-level rise and threatening coastal cities around the world.
- Polar bears, walruses and other ice-dependent marine mammals become extinct in the Arctic as the icecap disappears.
- Drought, fire and searing heat strike the Mediterranean basin.
- Declining snowfields threaten water supplies in California.
- A third of species worldwide face extinction as the climate changes.

Three degrees
- The Kalahari Desert spreads across Botswana, engulfing the capital in sand dunes, and driving millions of refugees out to surrounding countries.
- A permanent El Niño (see tip, right) grips the Pacific, causing weather chaos around the world, and drought in the Amazon.
- Water runs short in Perth, Sydney and other parts of Australia away from the far north and south.
- Agriculture shifts into the far north – Norway's growing season becomes like southern England is today. But with declines in the tropics and subtropics due to heat and drought, the world tips into net food deficit.

- The whole Amazonian ecosystem collapses in a conflagration of fire and destruction – desert and savannah eventually take over where the world's largest rainforest once stood. Huge amounts of carbon pour into the atmosphere, adding another degree to global warming.
- Hurricanes strike the tropics that are half a category stronger than today's, with higher wind speeds and rainfall. Wind speeds in the strongest storms could rise to 200 miles per hour.
- The Indus River runs dry due to glacial retreat in the Himalayas, forcing millions of refugees to flee Pakistan. Possible nuclear conflict with India over water supplies.

Four degrees
- Most of the Nile Delta is threatened by rising seas, as is a third of Bangladesh. Tens of millions more become climate refugees.
- West Antarctic ice sheet potentially collapses, pumping five metres of water into global sea levels.

Tip: El Niño is the name given to a reversal of currents in the east Pacific Ocean, which brings warm water to the coast of Peru, sparking rainfall in areas that are normally desert. In contrast, the lush forests of Papua New Guinea, where the rain usually falls, bake in drought. El Niño, in fact, changes weather across the globe.

- Southern Europe becomes like the Sahara, with deserts spreading in Spain and Portugal. People move north into temperate refuges in Scandinavia and the British Isles, which become increasingly overcrowded, resulting in further conflict.
- All glaciers disappear from the Alps, further reducing water supplies in central Europe.
- Permafrost melt in Siberia releases billions of tonnes of methane and carbon dioxide, meaning that global warming spirals upward.

Five degrees
- Earth hotter than at any time for 55 million years.
- Desert belts expand from the subtropics into temperate regions. Civilization collapses as humanity is unable to cope.
- Methane hydrate is released from underneath the oceans, sparking tsunamis in coastal regions and pushing global warming into an unstoppable spiral.
- Much of the world is uninhabitable.

Six degrees
- Mass extinction scenario: the end-Permian mass extinction 251 million years ago was associated with six degrees of warming, and wiped out 90% of life on earth. No one is sure what happened, but a combination of volcanic CO_2 releases and methane hydrates

may have been the cause. (This was much worse than the end-Cretaceous mass extinction, 65 million years ago, which wiped out the dinosaurs.)

- Huge firestorms sweep the planet as methane hydrate fireballs ignite.
- Seas turn anoxic (without oxygen) and release poisonous hydrogen sulphide.
- Humanity's very survival as a species is in question.

(*Note*: this list is a very potted summary of a book called *Six Degrees*, also by Mark Lynas, published in March 2007 by Fourth Estate.)

If none of this sounds very appealing, then you're reading the right book. This guide aims to show how you can help to avoid these disasters by reducing your personal contribution to global warming. The table on page 29 indicates the changes that are needed on a global scale to avoid each successive degree of global warming.

TWO DEGREES – THE TARGET

You might have noticed from the discussion above that there are certain 'tipping points', after which global warming could become unstoppable. Several of these could be reached if temperatures cross the two degrees threshold, because at this level of warming, greenhouse gas releases from the soils and forests could take on their own unstoppable momentum. That would leave humanity powerless to intervene as our planet began cooking. For this reason, many environmental groups and even governments (including the European Union) have begun to fix on the target of two degrees as a danger level that must not be crossed.

The graph on pages 20-1 shows how high temperatures might rise, in comparison to their levels over the last millennium. Note that the two degrees danger line is well below most of the scenarios for temperature change by 2100.

Staying below two degrees with any level of certainty would require global greenhouse gas emissions to peak by 2015 – quite a tall order, especially given the reluctance of many people and nations to face up to the scale and urgency of the problem. However, even if we overshoot this target, any cuts we do make will still help to slow the rate of change. At the very least, this gives human societies and natural species more time to adapt to rising sea levels, harsher weather and shifting climate zones. So it's not too late to act, but time is rapidly running out, and the era of procrastination is definitely over.

DEGREE CHANGE	ACTION NEEDED	CO_2 TARGET
One degree	Probably not possible	350 ppm (today's level is 380 ppm)
Two degrees	Cut global emissions by 2015	400 ppm
Three degrees	Cut global emissions by 2030	450 ppm
Four degrees	Cut global emissions by 2050	550 ppm
Five degrees	Allow constantly rising emissions	650 ppm
Six degrees	Allow very high emissions	800 ppm

WHAT IS A CARBON FOOTPRINT?

In short, a carbon footprint is a measure of an individual's contribution to global warming. Almost every energy-consuming activity we engage in has a carbon cost. Sometimes this carbon expenditure is obvious, such as the exhaust belching out of the back of a 'Chelsea tractor' 4x4 on the morning school run. At other times it is less obvious, like the 'food miles' inherent in imported produce we might buy at a supermarket. Your carbon footprint will be impossible to pin down exactly, but getting a general idea will help you to reduce your impact on the climate.

Many people (especially those who work in the fossil fuel industries) suggest that getting to grips with climate change will require terrible wartime-style sacrifices, where we all have to sit shivering around a single candle all winter while wearing five jumpers. But actually, reducing your carbon footprint is surprisingly easy, and not only does it not require hair-shirt sacrifices, it will probably increase your quality of life substantially. Cycling or walking rather than driving, eating local produce rather than supermarket junk and having some solar panels on your roof are hardly going to bring your world to an end. Sandals are, of course, optional.

Some comparisons between countries

We've all heard the cliché that Americans are greedy polluters, while Africans have a much lighter impact on the planet. But is it true? Different countries have radically different carbon impacts, depending on their pattern of development and the lifestyles of their citizens. America is in fact quite near the top of the global warming polluters' league table, with almost 20 tonnes of carbon dioxide emitted per person. But the United Arab Emirates and Qatar, with their oil-rich economies, have twice the per-person impact of the US.

European countries tend to have about half the per capita emissions of the United States: the UK, for example, has 9.4 tonnes to America's 19.8. So our carbon footprints are only half as large as those of the average American. But they are still big on a global scale – each Brit produces ten times as much CO_2 as each Vietnamese. Global warming naysayers are fond of pointing the finger at China and India, as if they are to blame for the problem. But China's emissions are only 3.2 tonnes per person while India's are 1.2 tonnes per person.

Africa is often forgotten altogether, and African per capita emissions are indeed minuscule in comparison to ours. Tanzania's people emit only 0.1 tonnes each – one hundredth of our per-person total – while Chad

Tip: At the moment it is not possible to say when each degree of temperature rise will be reached – because that depends on decisions we have all yet to make. If we continue to increase carbon emissions, we might reach two degrees as soon as 2030. But if we cut back our greenhouse gas emissions sharply, we may avoid two degrees altogether.

and Mali have emissions so low that they are not even measurable. It is interesting to note how African people are often blamed for environmental damage such as deforestation and the hunting of wildlife, but in global warming terms they are greener than green. The only exception is highly developed South Africa, which, with 7.8 tonnes per person, is nearly up to European standards. Anyone who has travelled around Johannesburg's traffic-choked motorways will quickly understand why.

Of course, because developing countries have such huge populations, their global impact is considerable. China is the second-largest emitter in the world (after the United States, of course), with 14.5% of the world total. India is number four in the list, with 5.1% of the total, while the UK is seventh, with 2.3%. Because of their large total impact and the rapid scale of their economic development, it is clearly true that global

---

Tip: The UK often claims to be doing most towards bringing down greenhouse gas emissions, and the government projects that we are on target under the Kyoto Protocol to meet our reduction of 12.5% by 2012. However, a tougher government target of 20% reductions by that date has now been quietly dropped, and emissions are currently rising again – thanks mostly to the general public driving and consuming more.

warming cannot be solved without engaging developing nations in the effort. But while such dramatic per capita disparities remain, China and India have a fair point in refusing to take the first steps.

Contraction and convergence

Aubrey Meyer of the London-based Global Commons Institute has proposed an international solution – 'Contraction and Convergence' (C&C). This proposal recognizes that the only realistic way to avoid global political stalemate is to accept the need for equity – that each person has an equal right to the use of the atmosphere. At the moment, the wide divergence in carbon footprints means that each American is using nearly twenty times as much atmospheric space as each Indian. By demanding that India makes cuts at the same time as the US, the American government is in effect proposing to cement this inequity – something the Indian government is understandably unwilling to sign up to.

C&C gets around this problem by putting in place a framework for 'convergence' to equity where, by a negotiable date (say, 2030), each country in the world will have an equal emissions entitlement based on its population. While all of our carbon footprints might not be the same by that date, our rights would be.

So people in rich countries who want to use more than their fair share would have to pay for the right to do so by buying unused allocations from people in poorer countries. The result would likely be a net transfer of wealth from rich to poor, which would help tackle global poverty at the same time as global warming. This wouldn't be charity, but trade – something world leaders are more likely to sign up to.

Of course, convergence is only one half of the equation. The other would be 'contraction', where global emissions contract downwards towards a sustainable level that would avoid serious climate change damages.

So far, C&C has gained substantial support from the African group of nations, while in the UK it has been recommended by the Royal Commission on Environmental Pollution and the Archbishop of Canterbury. Aubrey Meyer suggests that C&C could provide a stronger framework for climate change action once the Kyoto Protocol lapses in 2012.

Kyoto and international negotiations
In theory, the international community is already committed to solving climate change, and has been for more than a decade. The UN Framework

Convention on Climate Change was signed in 1992 at the Rio Earth Summit by leaders from 150 nations, including the first President George Bush on behalf of the United States. In its Article Two, it said the following:

> 'The ultimate objective of this Convention … is to achieve stabilization of greenhouse gas concentrations in the atmosphere at a level that would prevent dangerous anthropogenic interference with the climate system.'

Trouble was, no one defined 'dangerous'. Moreover, the Convention was only voluntary, without specific measures to actually enforce its stated objective. Nothing much happened as a result, and greenhouse gas emissions continued to rise after the UNFCCC was signed. In order to address this, the Kyoto Protocol was negotiated in Japan in 1997 to try to give the Convention some teeth. Industrialized countries agreed that they should take the first steps towards cutting emissions, since they were the highest per-capita polluters. Different nations took on different CO_2 reduction targets: the United States agreed to a cut of 7%, while the EU got a cut of 8%. Japan and Canada each took on 6%, while Australia negotiated itself an increase in emissions by claiming special status as a smaller country. In total, emissions from the industrialized world under Kyoto were supposed to fall by 5% by 2012.

Kyoto was finalized amid great celebrations in December 1997, and hailed as a triumph for global environmental protection. If it had worked, perhaps this book would not have needed to be written. Unfortunately things went somewhat downhill after all the delegates went home and the ink began to dry. First, no one did anything to cut emissions. Japanese, Canadian and US emissions continued to climb steeply, despite the countries' Kyoto targets. European emissions were more stable, largely because of the happy accident of economic collapse in post-Communist Eastern Europe, which dramatically curtailed emissions from the old factories and power stations of the Soviet bloc.

Kyoto's negotiating process continued at annual UN summits after 1997, but each successive meeting introduced new loopholes that served to weaken the treaty. Then in 2001, a further blow came when the second George Bush pulled the United States out of Kyoto, arguing that it would be too much of a strain on his country's economy (though many pointed at his administration's close links with the oil industry as a deeper reason for the hostility to Kyoto). The US played a further negative role by undermining the negotiations at the UN climate summits, and by trying to organize an alliance of other Kyoto-hostile powers.

It took seven long years before the Kyoto Protocol even came into force. It finally became legally binding in February 2005 when Russia belatedly ratified the treaty. Even so, the US and Australia remain on the outside, and Canada has said that it will not be able to meet its emissions target. Within the EU, Ireland and Spain are also grossly over-budget, and may have to buy 'emissions credits' from more thrifty countries.

CARBON RATIONING

Many people see rationing as the only verifiable long-term way to get people to change their lifestyles. It's the ultimate in carbon footprinting: rather than each individual having to work out their footprint using calculations like the ones later in this book, it would all be done automatically, perhaps with an electronic card. Each time you fill up the car, you would have to

---

Tip: It's not all bad news on the US front. Following the lead of Seattle's Mayor Greg Nickels, by September 2006, 295 city mayors, representing fifty million Americans, had pledged to implement the Kyoto agreement in their localities.

swipe your carbon ration card and surrender some units. You would also need enough carbon ration to cover your heating and other domestic needs.

The beauty of carbon rationing, according to its proponents, is that it would be government-enforced – making it much more likely to work than voluntary approaches. Because each person starts off the year with an equal ration, it would also be fair, just as everyone during the Second World War had an equal allocation of food and other necessities in order to jointly share in the war effort. One difference from wartime rations, however, is that carbon rations would be tradeable. This would introduce some flexibility – if you really *have* to have that power-boat, then you can buy an unused ration from someone who is happy to forgo such luxuries. The financial incentives would therefore reward those who kept their carbon emissions low, because they could sell their unused ration on the open market.

At the moment, things work the other way round – it is often cheaper to be energy-profligate. Budget

---

Tip: Find out more about carbon rationing in Mayer Hillman and Tina Fawcett's book *How we can save the planet*, published by Penguin in 2004.

airlines, for example, have made air travel more affordable than the train, despite the cost of the extra emissions on the climate (see Appendices A and B on pages 160–70). And buying 'green power' from renewable sources is more costly than simply sticking with the old emissions-heavy electricity from coal and gas-fired power stations. Carbon rationing would change all that.

THE GREAT ENERGY DEBATE

Nuclear power or wind farms? Tidal barrages or 'clean' coal? One of the most divisive issues in Britain today is energy. With gas and oil prices at historically high levels, energy is no longer too cheap to bother thinking about. Instead, it's on everyone's agenda, from the government on downwards.

Nuclear

The most controversial issue of all is probably nuclear power. It seemed to be on the wane after the 1986 Chernobyl disaster, but global warming may prove to have been nuclear's renaissance. Unlike the belching smokestacks that accompany coal or gas-fired power stations, nuclear fission releases no carbon dioxide into the atmosphere, so is officially carbon-neutral. Processing uranium ore and building the stations does have some greenhouse gas penalty, of course, but so does transporting coal and building a coal-fired generating plant. There is no doubt that nuclear's emissions are pretty low.

But that isn't what irks many anti-nuclear campaigners. Instead, they focus on the issues of radioactive waste and the dangers inherent in the fission process. Radioactive waste will stay deadly for thousands of years – what kind of legacy is that to leave for our great-great-grandchildren? There are also fears that terrorists could bomb reactors, releasing Chernobyl-style plumes of radioactive fallout onto nearby cities. Perhaps the strongest argument against nuclear power is that it is very costly – decommissioning the last generation of British reactors is already projected to cost £70 billion, making them the most expensive power source ever. If taxpayers' money is shovelled towards a new generation of nuclear stations, this will mean less support for renewables like wind and solar.

Wind

The second-most divisive energy source is probably wind. The arguments are dressed up in all sorts of technicalities, but let's face it – some people simply hate the sight of windmills. Why they don't get similarly annoyed at the legions of electricity pylons marching across our precious landscape, or the roaring ribbons of motorway tarmac which bisect some of our natural treasures, remains a mystery. Beauty – and ugliness – is in the eye of the beholder, and the debate about wind has much to do with aesthetics.

But there are some serious issues, too. One of the accusations levelled at wind proponents is that it is an intermittent energy source – sometimes the wind simply doesn't blow, and in that case, how do you keep the lights on? The answer is that the grid system needs backup-generating capacity, and that wind turbines need to be based in far-flung places so that all of them are highly unlikely to be becalmed at once. We'll discuss small-scale wind in more detail on pages 143–46, but for the moment it might be useful to remember that it could, in principle, supply more than 20% of the UK's electricity supply.

Coal

Coal supplies 35% of the UK's electricity, but at a much heavier CO_2 cost than gas. The UK still possesses relatively abundant coal deposits, but if these are burned it will further worsen our contribution to global warming. One possible interim solution is to pipe waste CO_2 underground and hope it stays there – a technique called 'geological sequestration' or 'carbon capture and storage'. This is already being carried out in the North Sea, where the Norwegian oil company Statoil injects carbon dioxide into a saline aquifer under the sea bed. One million tonnes of CO_2 per year are disposed of in this way, and the hope is that coal-burning power stations could install the same technology. However, the geology under existing power stations is not always suitable, and it remains much cheaper – in the absence of international legislation – simply to let the carbon go up the chimney. No power station yet employs carbon capture and storage technology.

Gas

More than 40% of Britain's electricity is generated using gas-fired power stations, making it the biggest single source. Gas releases CO_2 into the atmosphere, and is also getting more expensive as North Sea gas supplies run out. The 'dash for gas' meant that coal

---

Tip: One of the biggest concerns with gas is security of supply. North Sea gas reserves are already running out, so most of western Europe's supply comes from Russia. President Putin could leave us in the cold and dark at the flick of a switch. A dispute with Ukraine in January 2006 led to a reduction in the gas supply to France, Germany, Hungary and Italy, illustrating the Russian stranglehold over Europe's energy.

capacity was phased out in favour of gas during the late 1980s, reducing Britain's carbon emissions. This was hardly for environmental reasons, however – the Conservative government was mainly concerned to break the power of the miners' unions. But in 2005, with gas getting more expensive relative to coal, more coal was burned, making emissions rise again. Most coal burned in Britain is now imported from abroad, particularly from South Africa and Russia.

The former president of the Royal Society, Lord May, was only half joking when he said that the only truly acceptable form of energy generation to most people would be magic.

Energy facts for the UK

- Carbon dioxide emissions fell by 5.3% between 1990 and 2005, thanks largely to the power switch from coal to gas.
- In 2005, emissions rose by 0.3%, and the government announced that it would miss its domestic target of a 20% CO_2 reduction by 2010.
- Carbon dioxide accounts for 85% of the UK's greenhouse gas production, with methane and nitrous oxide contributing 7% and 6% respectively.
- In 2005, the UK emitted 567 million tonnes of carbon dioxide.

- Power stations are the largest single source of emissions. Drax power station in Yorkshire is the biggest, burning ten million tonnes of coal a year to produce 7% of the UK's energy, and emitting 20.8 million tonnes of CO_2 in the process.

Energy sector breakdown

The chart below shows the amount of energy used by different economic sectors of the UK. These statistics illustrate the important role individual choices play in reducing the UK's overall emissions. Add together the

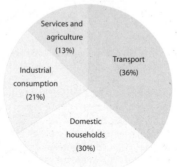

ENERGY SECTOR BREAK DOWN

Services and agriculture (13%)

Transport (36%)

Industrial consumption (21%)

Domestic households (30%)

Source: Department of Trade and Industry

transport and domestic sectors, and individuals can directly influence emissions from two-thirds of the UK's entire economy. That is why carbon counting is so important, because by becoming more aware of the carbon impacts of each consumption decision, people can quickly reduce their role in causing global warming.

WHAT IF GLOBAL WARMING IS A 'HOAX'?

The US Republican senator James Inhofe has called global warming 'the greatest hoax ever perpetrated on the American people', while the novelist Michael Crichton has penned a thriller that paints environmentalists as terrorists inventing global warming to satiate their lust for money and power. (Crichton was invited by Inhofe to testify to the US Senate and talk more about this fantasy as a result.)

Global warming is nothing if not controversial, and a dedicated group of people who call themselves 'sceptics' spend a lot of time challenging the mainstream scientific consensus. Many of these sceptics – or climate change 'deniers', as they have been termed – receive funding from the oil and coal industries, so their claims should be taken with a pinch of salt. But their arguments are just as likely to surface in newspaper columns or at the pub, so it is worth looking at a few of them.

Argument no. 1: 'In the 1970s, scientists were predicting an imminent ice age. Why should we believe them now?'

Response: No, they weren't. There were indeed one or two scientists who suggested the cooling trend between the 1940s and the 1970s could be the start of the descent into an ice age, but this was a fringe view, not a mainstream one. The current consensus on global warming is probably unprecedented for any scientific subject, and consequently ought to be taken seriously.

Argument no. 2: 'The cooling trend between the 1940s and the 1970s, when greenhouse gases were rising, shows that CO_2 doesn't cause global warming.'

Response: Wrong again. Remember that CO_2 isn't the only factor in this equation. During this period, emissions of sulphate aerosols – which cool the climate – were very high, and natural variability may also have played a role.

Argument no. 3: 'The warming trend that we've seen on the planet is a result of all these cities, which cause an "urban heat island" effect. You didn't think of that, did you?'

Response: Sorry, but yes we did. Analyses of global temperature trends are corrected in order to remove any signal from urban heat islands.

Argument no. 4: 'Volcanoes release more CO_2 than does all of humanity, so why do we always get the blame?'

Response: Over long periods volcanoes do indeed vent carbon dioxide. But on a year-to-year basis this impact is negligible – less than 2% of human carbon emissions.

Argument no. 5: 'Global warming theory depends on computer models whose validity is questionable.'

Response: Everything in science is questionable. But climate models are actually very good at simulating past climate change over the previous century, suggesting that they might also have something to tell us about the next. It is also notable that models can't reproduce the current warming trend if the rise in CO_2 is left out.

Argument no. 6: 'Weather forecasters can't even predict what is going to happen in three days, so why should we believe their forecasts for 100 years?'

Response: This is to confuse weather and climate.

Weather is very chaotic and unpredictable, but climate refers to averages and is therefore much more predictable. It is because climate is predictable that you can say that July 20 will almost certainly be warmer than January 20 – even if no one can tell whether it will rain on either of those two days.

Argument no. 7: 'Other factors could be causing the warming, like the sun. Perhaps humans aren't to blame.'

Response: Solar variations can indeed affect the weather, but solar output is not changing significantly enough to be able to account for current global warming. In any case, other explanations aren't needed, because the change in greenhouse gas concentrations explains the current warming trend perfectly well, however inconvenient this might be for some people.

Argument no. 8: 'Environmentalists are Communists, trying to undermine our society, and that's why they have invented global warming.'

Response: Did you take your pills this morning?

USING THIS BOOK

While the big energy decisions are taken by governments, we individuals still have a crucial part to play in saving the planet from the threat of global warming. Indeed, the role of the public is particularly important given the failure of politicians so far to give a clear direction.

Even though the government refuses to take the climate issue seriously enough to impose carbon rationing, there is nothing to stop us, as individuals, from rationing ourselves in order to get our carbon footprints down to tolerable levels. That's where this book can help.

Carbon emissions per person in the UK are currently 9.4 tonnes of CO_2 per year. While only half the level of American emissions per person, this is still ten times what Indians get by on. A sustainable per capita carbon emissions level, as section III will show, would be about one tonne – implying a 90% cut for us all.

Each section of this book will lay out your options for getting towards this target. We'll begin by calculating your current carbon emissions in section II, and explaining how you could make savings in each area, from cooking to car driving. We'll calculate your grand total in section III, and then explore how you can generate your own renewable energy at home to bring your emissions down in future years.

Don't forget – as well as changing your own life, it is important to help other people change theirs. Don't be embarrassed to interrupt genteel dinner parties with heartfelt pleas to guests not to take their accustomed foreign holidays. Explain why you feel that climate change is the most important issue for our future, and why you want to play your part in solving it. Give them a copy of this book, and help build the growing movement to take on the scourge of global warming.

II
Your carbon footprint

HOUSEHOLD EMISSIONS

GAS

How did we ever live without gas? Every time you ignite your gas cooker or step into a gas boiler-heated hot bath, think how different life once was without it. We have become used to gas as a domestic necessity, and are now highly dependent on this non-renewable resource.

However, gas also produces carbon dioxide when you burn it. That boiler exhaust vent high up on your outside wall is pumping CO_2 into the atmosphere every time you switch on your hot tap.

Space heating and hot water

Heating is the biggest single user of energy in domestic households, soaking up 70% of the energy we use. The European Union calculates that four billion tonnes of CO_2 each year – a whopping 14% of the EU's total greenhouse gas emissions – comes from heating.

Tip: A national network of over fifty energy efficiency advice centres can provide all sorts of advice and information about how you can save energy in your home, tailored to your own personal circumstances. The freephone number is 0800 512 012. They're waiting for your call.

This means that reducing the emissions associated with space heating is almost certainly the single most important thing you can do to reduce your carbon footprint – far more important than sexier things like installing solar panels or driving a hybrid car.

First, turn down your thermostat. Reduce the temperature of your home by just 1ºC, and save 300kg of CO_2 per household per year. Yes, that might mean putting on another jumper. (Thank goodness knitwear is coming back into fashion.)

Insulation

Unglamorous it may be, but probably the most important thing you can do in your house to save on energy is to insulate it properly. Heat pours out from millions of poorly insulated roofs every winter – for an easy visual guide, note how snow and frost will lie all day on some roofs but melt quickly on others. That's because of differences in insulation – with little or no loft insulation, up to a third of the householder's

Tip: Many studies have shown that a majority of people do not fully understand the heating and hot water controls on their boiler and central heating systems. Get to grips with all the settings, and make sure that your hot water isn't too hot (below 60ºC) and that your central heating isn't on longer than it needs to be.

heating is going towards warming the air above their house. Not a good idea.

All lofts should be insulated to a minimum depth of 270mm; anything over this will continue to add to the benefits. You can do this yourself by buying rolls of insulation and unrolling them over the rafters (tip: don't walk between the rafters or you may fall through the ceiling).

The cheapest option is to buy fibreglass or rockwool rolls, but these are unpleasant to work with because the fibres get everywhere and itch. A pleasanter alternative may be sheep's wool, which feels and smells just like a big woolly jumper. It's rather more expensive, but if there's any left over you can sleep in it. There is also insulation made from recycled newspapers, known as Warmcel, which you can buy in convenient 8kg bags.

Contacts: For sheep's wool insulation, known as Thermafleece, see www.secondnatureuk.com. For Warmcel, visit www.excelfibre.com.

Tip: Gas fires are a terrible waste of heat – most of it goes up the chimney. Open wood fires are also extremely inefficient. Get a wood stove, or rely on central heating via radiators.

Another absolute must is cavity-wall insulation. Any brick house built after the 1920s should have a gap between the two layers of brick in the walls, which needs to be filled up with insulation. It is a scandal that millions of homes in the UK still waste huge quantities of heat by not doing this simple process. It's both easy and cheap – installers simply drill holes in the outside of your walls and inject in the insulating material, usually polystyrene beads or mineral wool. According to the Energy Saving Trust, payback times for cavity-wall insulation can be as little as two years.

To get a list of installers in your area, contact your local energy advice centre. You can find out where they are by entering your postcode on www.est.org.uk/myhome/localadvice.

You can tell if you have cavity walls by checking the brickwork pattern on the outside of your house (see page 57). In cavity walls, the bricks are all evenly laid end to end, whereas solid walls have a half-brick pattern in between every full length one.

Tip: Subsidized cavity wall insulation is available throughout the UK. Getting this serious problem fixed may cost you as little as £150, and may be free for pensioners and those claiming benefits.

You can also check by measuring the width of your walls at a doorway or window. Cavity walls will be at least 30cm thick, whereas solid walls are only about 23cm thick.

If your home was built after 1995, its cavity is likely to have been insulated during construction, so you can probably afford to relax about this one.

If you don't have cavity walls – if you live in a Victorian house, for example – don't despair. You are not alone: out of the UK's housing stock of 24.5 million dwellings, 36% are built with solid walls. Solid walls do tend to lose heat quickly, so they need insulation if heating bills (and emissions) are not to soar to stratospheric heights. According to the National Insulation Association, 33% of domestic heat can be lost through uninsulated walls. There are two basic options: external and internal wall insulation.

External solid wall insulation
The first thing to say about this is that it's a big job, but is worth considering especially if you are already undertaking major renovations. It involves attaching slabs of insulation material to the walls outside your home, and then coating it all with a render or cladding finish.

SOLID BRICK WALL

CAVITY WALL

Advantages: It will keep the insides of your home at a more even temperature, and help to prevent condensation, without taking up valuable space inside rooms. It could also save a quarter of your annual fuel bill, and reduce CO_2 emissions by 700–1000kg per year.

Disadvantages: Tends to be very costly, at around £45 to £65 per square metre applied – adding up to between £3,500 to £5,500 for an average semi. Get several quotes. To find out more from a registered installer, visit: www.inca-ltd.org.uk.

Internal solid wall insulation

This involves putting up insulation around the interior of a room, either with ready-made boards (like plasterboard, but with an insulating layer), or by making a wooden frame and stuffing fibre insulation into it. The latter can work well on old stone houses with uneven walls, because you end up with straight edges in your house. Make sure your home is properly damp-proofed before considering internal insulation, or things will just get worse. You'll also need to put up a polythene membrane to stop condensation.

Advantages: Cheaper than external insulation, and a serious option for experienced DIYers, particularly if you are already doing internal renovations. Your house will heat up much more quickly and stay warmer for longer periods. Boards or flexible thermal lining cost between £35 and £40 per square metre.

Disadvantages: You lose living space equivalent to the width of insulation installed. To be effective, boards need between 6cm and 10cm of insulating layer.

To find out more or to contact a registered installer, visit www.nationalinsulationassociation.org.uk/house holder/householder-nia.html.

Floor insulation

This may also be a good option if you have draughts coming up from between your floorboards – they can

AVERAGE COSTS AND SAVINGS FROM TYPICAL ENERGY EFFICIENCY IMPROVEMENTS

Measure	Annual saving (£/yr)	Installed cost (£)	Installed payback (years)	DIY cost (£)	DIY payback (years)
Cavity-wall insulation	£100–£120	£135	1–2	–	–
Internal wall insulation	£210–£260	>£40/m²	–	–	–
External wall insulation	£220–£270	>£1800	>7	–	–
Loft insulation (0–270mm)	£140–£170	£135	<1	>£275	1–2
Loft insulation top up (50–270mm)	£40–£50	£200	4–5	>£200	4–5
Draught proofing	£10–£20	>£75	~7	>£50	3–5
Floor insulation	£30–£40	–	–	>£100	>3
Filling gaps between floor and skirting board	£10	–	–	£20	2
Hot water tank jacket	£15	>£5	<1	>£10	<1
Primary pipe work insulation	£10	–	–	£5–£10	1–2

Source: Energy Savings Trust

be filled with a sealant, or by laying an insulating layer underneath the boards. You can also reduce these draughts by laying carpets or rugs down – but don't put down a carpet if you have under-floor heating, as it will stop it from working effectively. You may also be able to add insulation under a raised ground floor.

There are grants available for installing all types of insulation, particularly if your house currently has none at all. Grants are paid directly to a professional installer, so all you have to fork out is the difference. If you are on benefits, chances are good that putting in insulation will be completely free. To find out all the grant options, visit the Heat Project on www.heca.co.uk.

Draught-proofing

Even less sexy than insulation – but again, it can save you both money and CO_2 emissions to install it properly. The average home loses 20% of its heat through ventilation gaps and draughty windows. Don't forget, though, that adequate ventilation is essential,

---

Tip: Never heat your conservatory! Conservatories are great for spring, summer and autumn use – and will help keep the rest of your house warm in sunny weather – but are never well-enough insulated to retain heat well once the sun goes down, so don't use them as winter-time living space.

especially if you have a solid-fuel stove or a gas fire, and ventilation is also needed in the kitchen and in bathrooms. Probably the easiest way to draught-proof windows is to hang curtains in front of them and close them at dusk. Makes things feel cosier too. Huge amounts of heat are also lost through draughty or single-glazed windows. Double-glazing can be an expensive upfront investment, but will pay off in the longer term – the best wooden-framed double-glazed units (which are filled with argon gas) can reduce heat loss from a building by 70%.

Your boiler

Check what kind of boiler you have, and how efficient it is. If it is more than ten years old, chances are that you would save money and emissions by upgrading it to a far more efficient condensing boiler. According to the Energy Savings Trust, this could straight away knock 40% – or between £190 and £240 per year – off the average household gas bill.

If everyone in the UK with gas central heating installed a high efficiency condensing boiler, enough energy would be saved to heat 3.7 million homes for a year. If you are installing a new kitchen or bathroom, then why not install the latest high-efficiency condensing boiler at the same time?

Since early 2005 all new boilers installed have to be high-efficiency condensing boilers by law. The best models nowadays are up to 92% efficient – they work by recycling some of the heat in the flue gases so it isn't wasted.

There are various grants available for improving the energy efficiency of your house – the government operates a scheme called 'Warm Front' (with variants 'Warm Homes' in Northern Ireland, 'Warm Deal' in Scotland and 'Home Energy Efficiency Scheme' in Wales). Some energy suppliers and local authorities also provide grants. To find out if you qualify, visit www.est.org.uk/myhome/gid.

As well as your heating system, your boiler will also be responsible for supplying you with hot water. There are various different systems: a combination boiler fires up on demand when you switch on the hot tap (and also supplies hot water to your central heating), while a stored system will use a boiler to heat water in a tank. If you have a hot water tank it may also be heated electrically, via an immersion coil.

Tip: To find out how efficient your current boiler is, visit the www.boilers.org.uk database and search for the manufacturer and model number.

SAVING ENERGY

When you are shopping for new appliances, look for the energy saving recommended logo. It's your guarantee that the product will save energy, cost less to run and help the environment.

Certification mark

A hot water tank should be insulated with a 'jacket' (which should be nice and cosy, at least 75mm thick), easily obtainable from your local DIY store. It will cost as little as £10 and will save £20 a year in your annual fuel bills. Hot water pipes should also be lagged with insulation – again, this is an easy job to do yourself. You can buy lagging from all DIY shops, but don't forget to measure the diameter of your pipes first.

Always ensure that anyone who works on any gas-related appliance is CORGI registered. To find a professional installer near you or to check someone's ID registration, visit www.corgi-gas-safety.com.

Tip: Rather than buying a special jacket for your hot water tank, just pack around it with an old duvet – just as effective and much cheaper.

You can also search for the most efficient boiler (or any other appliance) via www.est.org.uk/myhome/ efficientproducts/recommended.

Keeping cool

Air conditioning soaks up enormous amounts of energy, while electric fans are also fairly power-hungry (though much less so than air-con). Here are some tips for energy-savers wanting to keep cool in the summer months:

- Insulation again. As well as keeping your house warm in the winter, it also keeps you cool in the summer by stopping your home from responding too quickly to changes in outdoor temperatures.
- Keep blinds or curtains drawn across south-facing glass windows.
- Try to generate a through-draft by keeping windows open on opposite sides of the house.
- Heat rises, so if you have a skylight or a window open at the top of your house and let air in at the bottom through doors or windows you can also generate a through-flow of cooling air.
- Turn off appliances that might be heating things up further, like computer monitors and the oven.
- Air-dry dishes rather than using a dishwasher: the evaporating water will cool your kitchen.

- Plant trees to shade big windows or over-hot conservatories, and have vines climbing up the outside of your house on trellises to keep things cool and shady.
- Slow down: the hottest days are also the ones you want to avoid working out in your personal gym…

Calculate your gas carbon footprint

First, you need to dig out your quarterly gas bills. Your actual gas meter will be counting units, but we need the figures for kWh, which will also be on the bill.

Gas use (kWh)

First quarter   ....

Second quarter   ....

Third quarter   ....

Fourth quarter   ....

Total for the year   ....

If you really can't find your gas bills, you could just use these bog-standard averages, for the average house, inhabited by average people in an average place.

Small house:	10,000 kWh per year
Medium house:	20,500 kWh/yr
Mansion:	28,000 kWh/yr

To convert this into carbon emissions, multiply by 0.19, and to get your personal total, divide by the number of adults in the house.

Your total gas carbon footprint for the year

.... kilograms

Make a note of this figure. You'll need it later, when we come to tot up the overall total.

Your oil, coal and wood footprint

If you use oil for heating instead of gas (you're highly unlikely to have both), enter the number of litres you use in a year.

....

Multiply this figure by 2.975.

Tip: Patio heaters have increased Britain's greenhouse gas emissions by 380,000,000 tonnes a year. The average patio heater, when used for two hours, produces as much CO_2 as a car does in a day.

Your total heating oil carbon footprint for the year

.... kilograms

If you are heating your house using wood (preferably locally cut from a sustainable source, and burned in a clean-burning stove) then you can put down a zero for your space heating emissions. As trees grow, they re-absorb the emissions from the wood you burn, instead of making a net addition of CO_2 to the atmosphere, as fossil fuel burning does.

If you are burning coal, on the other hand – especially if it's in an open fire – emissions will be very high: this is the worst-possible heating option. This is because coal is the most carbon-intensive fossil fuel: in terms of heat delivered when burned, it generates much more carbon dioxide. Take your total weight of coal burned in kilograms and double it to get the CO_2 impact. For both coal and oil, divide the total by the number of adults in the house to get your personal contribution.

Your total coal carbon footprint for the year

.... kilograms

ELECTRICITY

Despite numerous polls showing that public concern over climate change is rising, carbon emissions from the domestic sector are rising too, as people pile more and more consumer gadgets into their houses. Although households are becoming steadily more efficient in how they use energy (the average fridge, for example, uses much less energy than an equivalent machine did thirty years ago), this improvement is being outstripped by our insatiable demand for more and more electronic goods. Between 1972 and 2002, electricity used in the household sector doubled, and is projected to rise

TYPICAL ENERGY-CONSUMING PRODUCTS IN A 1970'S HOME*		
Television	Electric kettle	Cooker
Vacuum cleaner	Washing machine	Cassette player
Electric bar heater	Iron	Fridge
	Electric blanket	DIY appliances
Hi-fi music system	Radio	Toaster
Hairdryer	Sewing machine	Lamps

*Source: Energy Savings Trust report – 'The rise of the machines', June 2006. www.est.org.uk

TYPICAL ENERGY-CONSUMING PRODUCTS IN A 2000'S HOME*

Televisions	Digital radio	Portable fan
Video players	Mini hi-fi systems	Vacuum cleaner
DVD player/recorder	Washing machine	Computer
Portable music players	Tumble dryer	Monitor
Mobile phones	Dishwasher	Printer
Hairdryers	Playstation/games console	Scanner/fax
Hair irons	Cappuccino maker	Digital camera
Electric toothbrushes	Digital clock/radios	Set-top box
Wireless phone/answering machine	Electric lawnmower	Electric shaver
Slave portable phone handsets	Strimmer	Steam iron
Electric kettle	Microwave	Juicer
Smoothie maker	Electric oven	Home security system
Magimix	Electric hob	Broadband/wi-fi
Ice-cream maker	Extractor fan	Halogen light-bulbs
	Large fridge/freezer	Personal care products
	Drinks cooler	Power tools
		Electric blanket

another 12% by 2010. We all want more and more. The boxes on pages 68–9 show the difference between how we live now and how we lived three decades ago. Note the different lengths of the lists.

You will notice that many of these gadgets use power even when they are not being used – the notorious 'standby'. If televisions and video players are operated by remote control, they may never be properly switched off. Phone chargers continue soaking up power if they are not unplugged or switched off at the plug. That's why they still feel warm.

The rise in single-occupancy households also means that more energy is used per capita. You still need to heat a whole house for one person, just as you do for two or three, for example. In 1971 there were around three million people in England living on their own. By 2005 this number had more than doubled to nearly seven million, and official projections suggest that it will continue rising to around nine million by 2021.

Tip: Britons waste around two power stations' worth of electricity each year just by leaving TV sets and other electronic items on standby. The average household has 12 gadgets left on standby or charging at any one time.

Add in population growth, and that gives the UK another two million households by 2020 – each one of which will want all the same mod cons as those listed on page 69. According to research by Dr Jo Williams of University College, London, one-person households consume 38% more products, 42% more packaging, 55% more electricity and 61% more gas per capita than four-person households. They also produce more than half a tonne of extra waste per year.

In addition, single-person households almost by definition produce more carbon dioxide per person than multiple-occupancy ones. But not all are equal – pensioners tend to be more thrifty, for example. Lifestyles are critical; there can be huge differences between individual households based on how they choose to live their lives.

Living alone also has resource-use implications outside the immediate household. Much of the demand for new housing, with all that this entails

---

Tip: Consider buying an electrisave, a gadget that slots onto the wires under your meter and keeps a running total of electricity use. That way you can see what is soaking up the most juice. Find out more at www.electrisave.co.uk.

in terms of energy use for building materials, comes from the fact that more people are taking up entire houses by themselves.

So one of the best things you can do to reduce your carbon footprint is to share it with someone else. That doesn't necessarily mean shacking up with someone in the non-platonic sense, if you haven't yet found the right man or woman – co-habiting is something that many people do to reduce the bills, whether they are students or professionals. If you live alone, simply getting a lodger will almost halve your personal domestic carbon footprint at a stroke, because you will share heating and electricity.

Working out your electrical carbon footprint

Dig out your electricity bills – they are normally issued quarterly. If you can't find any, contact your electricity supplier, who should be able to quote you a figure over the phone. If your juice comes from a renewable source (more of that on page 74), enter 0 in all the boxes, sit back and feel the love.

Tip: It might be easier to take the total electricity bill for the year and divide it by the unit cost per kWh (about 11p at the time of writing) to get your total electricity use.

Electricity use (in kWh)

First quarter ....

Second quarter ....

Third quarter ....

Fourth quarter ....

Total for the year ....

If you can't find your electrical usage, just use the following averages:

Small house: 1,650 kWh per year
Medium house: 3,300 kWh per year
Mansion: 5,000 kWh per year

To convert this into carbon emissions, multiply by 0.43, and to get your personal total, divide by the number of adults in the house.

Your total electrical carbon footprint for the year

.... kilograms

Counting electrical carbon

If you're shocked at how high your carbon figures are, don't worry – there's a lot you can do to bring them down. The easiest step of all is simply to switch to a renewable electricity supplier. This is a company that matches your electricity use with power from wind turbines, small hydro-electric plants, solar power and other non-carbon sources. It's very easy to switch – you can do it online, or with a phone call – but you will need to have an old bill to hand in order to be able to quote the 'electricity supply number' from it.

The exact 'green tariffs' available will depend on your location in the UK. What's on offer also varies, as does the price. Not all companies offer the same deal. Good Energy, for example, guarantees 100% renewable power, and is therefore probably the greenest option. Others may simply put the extra tariff money into a fund for renewables projects.

The table opposite gives you some guidance here, but all aspects are subject to change, so check online for the latest.

Contact: Go to the website www.greenelectricity.org to find out what's available in your area, to compare prices and to follow sign-up links.

Company and tariff:	Ecotricity Ltd – Ecotricity
What it offers:	Small independent company. Your cash helps them to invest in renewable energy. Tops the list in terms of amount spent per customer on building new renewable sources.
Where it is available:	England, Wales and Scotland
Company and tariff:	Good Energy Ltd – Good Energy
What it offers:	Guarantees 100% renewable energy to customers. Also supports home-based and independent generators.
Where it is available:	England, Wales and Scotland
Company and tariff:	Green Energy Plc – Green Energy
What it offers:	Aims to purchase most of its electricity from renewable sources, and offers chance for customers to become shareholders in the company.
Where it is available	England, Wales and Scotland

Company and tariff: EDF Energy (London Energy/
SWEB/Seeboard) – Green tariff

What it offers: Costs about £13 a year more; this
money goes into a fund and is
used to support renewables
projects. You are also given two
low-energy lightbulbs. EDF is
French and generates most energy
from nuclear power.

Where it is available: England and Wales

Company and tariff: RWE npower - Juice

What it offers: No premium, and npower donates
£10 per customer into a fund to
support renewables projects.
Developed in partnership with
Greenpeace, though parent
company RWE has hugely
polluting brown coal interests in
Germany. Electricity is from North
Hoyle wind farm.

Where it is available: England, Wales and Scotland

Company and tariff: Powergen – GreenPlan

What it offers: Premium of around £18 paid by
customers; this is matched by

Powergen and used to fund community renewables projects. In addition, WWF gets £3 for every new customer. Parent company Powergen is the UK's largest CO_2 emitter, however.

Where it is available: England, Wales and Scotland

Company and tariff: Scottish Power – Green Energy H_2O

What it offers: No premium for this tariff, but electricity is supplied from existing hydro-electric plants, so cannot be considered an additional CO_2 reduction.

Where it is available: England, Wales and Scotland

Company and tariff: Scottish Power – Green Energy Fund

What it offers: Doesn't directly buy renewable electricity but invests £15 a year per customer (matched by Scottish Power) in renewables schemes.

Where it is available: England, Wales and Scotland

Company and tariff: Scottish & Southern – RSPB Energy

What it offers: Small premium of 3% above cost from your local supplier. 100% renewable, but 90% comes from large hydro. Money also goes to the RSPB for each customer.

Where it is available: England, Wales and Scotland

Company and tariff: Northern Ireland Electricity – Eco Energy

What it offers: No premium, but demand needs to rise for additional renewables capacity to be built. Also contributes to a fund for community renewables projects.

Where it is available: Northern Ireland

Tip: Switch your preferred washing machine cycle from 60°C to 40°C, saving 40% on the energy used. Dishwashers should be set on 55°C rather than 65°C. Washing up by hand is probably more economical – so long as you don't leave the hot tap running the whole time…

Green power will typically be slightly more expensive than the traditional dirty stuff. But once you're signed up, you can use as much electricity as you like, with a clear conscience! (Though that's not really the point.)

Reducing your electricity use

Stop using so much stuff! Do you really need a plasma screen TV? What was wrong with the old one? If you can't be persuaded to part with any of your gadgets, you can at least make sure that any new consumer electrical goods you buy have the highest energy efficiency rating.

On pages 80–3 there are tables listing the energy efficiency of major brands of four common household items. Obey the golden rule: anything less than A – avoid it.

If you want to check up on other models of electrical products, visit the Energy Saving Trust website, which can also suggest suppliers. Go to www.est.org.uk/myhome/efficientproducts.

Tip: Always let food leftovers cool down fully before putting them in the fridge. Defrost the freezer regularly. When cooking, put lids on pans – things will boil more quickly and less energy is wasted.

REFRIGERATORS	
Brand	**Energy efficiency class**
Bosch KS38R495GB	A+
Bosch KTR16424GB	A+
Bosch KTR16A20GB	A+
Bosch KTR18425GB	A+
Electrolux AEG SK91200-4i	A+
Electrolux AEG SK98840	A+
Electrolux AEG SU96040-4i	A+
Electrolux AEG SU99040	A+
Hotpoint HDR1901	A+
Hotpoint RLA56	A+
Hotpoint RLM36	A+
Indesit SAAN 300	A+
Miele K 854i-2	A+
Miele K 856i-1	A+
Miele K2319S	A++
Miele K2329S	A++
Siemens KT18RP90GB	A+
Siemens KT18RP90GB	A+
Vestfrost FZ 277 R	A+
Vestfrost FZ 349 R	A+
Whirlpool ARC 0850	A+
Zanussi ZERC0750	A+

WASHING MACHINES	
Brand	**Energy efficiency class**
Beko WMA665	AAA
Beko WMA765	AAA
Bosch WFR3269	AAA
Candy CM2166	AAA
Electrolux AEG L86810	AAA
Electrolux AEG LL1620	AAA
Hoover HNL6156S	A+AA
Hoover HNL9146Z	A+AA
Hotpoint BS1400	AAA
Indesit WIE 167	AAA
John Lewis JLWM1405	AAA
John Lewis JLWM1603	AAA
LG WM-14316FD	AAA
Miele W 2659 i WPM	A+AA
Miele W 2888 WPS	A+AA
Miele W 3622 WPS	A+AA
Servis M6006	AAA
Siemens WXLP166A	AAA
Smeg WMF16AX	AAA
Whirlpool AWO 10961	AAA
Zanussi ZWF1621 W/S	AAA
Zanussi ZWF1651 W	AAA

DISHWASHERS	
Brand	**Energy efficiency class**
Beko DWD5410	AAA
Beko DWD8650	AAA
Bosch SGS46E08	AAA
Bosch SRV43M03	AAA
Candy CDF312	AAA
Candy CDI PLAN	AAA
Electrolux F50860-M	AAA
Electrolux F65080VIL	AAA
Electrolux F86070VI	AAA
Hoover HND615	AAA
Hoover HND925	AAA
Hotpoint BFI680	AAA
Hotpoint SDW85	AAA
John Lewis JLBIDW1200	AAA
John Lewis JLDWW905	AAA
Miele G 1470 SCVi	AAA
Miele G 2670 SCVi	AAA
Neff S54T09X1	AAA
Siemens SF25T552	AAA
Smeg DF6FABNE	AAA
Whirlpool ADG 550	AAA
Zanussi ZDF601	AAA

TUMBLE DRYERS	
Brand	**Energy efficiency class**
Beko DRCS66	C
Beko DRVS62	C
Bosch WTA4108GB	C
Bosch WTE8630S	C
Bosch WTS84507	B
Bosch WTS86516	C
Electrolux AEG T56830	C
Electrolux AEG T57830	B
Electrolux AEG T59800	A
Hoover HNC 382T	B
Hoover HNC375T	C
Hotpoint CTD80	C
John Lewis JLTDC01	A
John Lewis JLTDC06	C
Siemens WT46E386GB	C
Siemens WT46E387GB	C
Siemens WT46S598GB	C
Whirlpool AWZ 341	C
Whirlpool AWZ 8477	B
White Knight CL 847 A Class	A
Zanussi TCE7227W	B
Zanussi ZBD5377W	B

Source: Energy Savings Trust.
www.est.org.uk/myhome/efficientproducts/recommended

Note that with tumble dryers, C is as good as it gets for many of them. The simple fact is, dryers just use a whole bunch of energy. Substitute a washing line, if at all possible; it's also better for your clothes.

Note also that the amount of energy used by an item depends on what you do with it. For washing machines, try to use low-temperature washes and keep fridge thermostats at a reasonable level. Use half-load or economy settings if dishwashers and washing machines are only half full. If you're really canny, you can even add extra insulation to the outside of your fridge to help keep it cold (but don't block up the area at the back where it vents out heat, or insulate around chest freezers, which lose waste heat through their jackets).

Lighting
Something you may have already heard a lot about is low-energy lightbulbs. 'The planet is dying – change your lightbulbs!' You know the sort of thing. Actually, conventional incandescent light bulbs are surprisingly wasteful – 80% of the energy they use goes on heat rather than light. Halogen lightbulbs – like the spotlights found in many kitchens – are even worse. Swap them for low-energy bulbs and you will also save money: they last on average twelve times longer

than old bulbs, and each one can knock £9 off your annual electricity bill.

In the old days, low-energy fluorescent bulbs used to be ugly and chunky and take hours to heat up with a sort of sickly bluish light. Luckily the march of progress has meant that they are now smaller and swankier than the traditional energy-wasteful alternative. And if you must have one of those annoying security lights switching on and off outside your house every time a cat walks by at night, at least fit a low-energy alternative bulb.

Standby

Simple problem, simple solution. Just switch off everything at the wall. Contrary to popular opinion, digital TV set-top boxes can (and should) be switched off. Whatever it is, if it needs to be on constantly, get rid of it, and ring up the manufacturers to berate them about their lack of climate change awareness (using their freephone number).

There are also, believe it or not, gadgets to reduce your use of gadgets. The Intelliplug switches off peripheral computer equipment without you having to know about it. (Though it doesn't work with laptops.) See www.oneclickpower.com.

TRANSPORT EMISSIONS

CARS

Every good environmentalist hates cars. Most greens still run them, however, because for all their ecological drawbacks cars are highly convenient – especially in a society that has been largely designed around high mobility and personal transport. Before cars were in widespread use, shops and amenities tended to be within walking or cycling distance. Now, with suburban sprawl and the rise of out-of-town shopping, people may have to drive miles to get a pint of milk or a loaf of bread, and can barely imagine life without their wheels. The average shopping trip in Britain is currently 4.3 miles – hardly a walkable distance.

Having said that, over a quarter of car journeys are under two miles in length – a distance that could easily be covered by bicycle. But transport in Britain has become subject to a vicious circle. The so-called 'school run' is a classic example. Whereas in the past children would walk or cycle to school, nearly 40% of primary school children are now driven to school in cars. One of the reasons cited by parents for this switch is that the roads are no longer safe for children to walk along because of the rising volume of traffic!

Despite the constant moaning of the car lobby on the pages of the *Daily Mail*, the real cost of motoring has actually declined since the Labour government came to power in 1997. Between that year and 2005 the real cost of motoring fell by 9%, whereas bus and coach fares increased by 15% and rail fares by 5%.

By itself, road transport accounts for fully a fifth of the UK's entire carbon emissions – totalling 33 million tonnes in 2004. Emissions from lorries have risen even faster than those from private cars, spiralling upwards by 25% between 1990 and 2004. If you think there are more lorries on the roads now than ever before, you're right.

But the government is in danger of killing motoring with kindness. Road traffic in the UK is on an unrelenting upwards trend, and has increased by 10% since 1997. The result is congestion as millions more people pile onto the packed-out roads of this small and crowded land.

The government now spends £1 billion each year widening existing roads and building new ones, but providing new road space simply tempts more people into making more car-based journeys, worsening traffic growth.

Much of the environmental lobby's bile has been focused on 4X4s, the fuel-inefficient monsters so beloved of middle-class parents living in affluent places like Chelsea – hence the nickname 'Chelsea tractors'. As well as being status symbols, 4 x4s confer a feeling of safety on their drivers, who know that they are unlikely to come off worst in an accident. This makes them more dangerous for everyone else, however – particularly pedestrians, more of whom would be killed in a collision with a 4x4 than with a smaller car.

A campaign group called the Alliance Against Urban 4x4s has been formed, whose members distribute fake parking tickets to offending vehicles. On the customary black and yellow background, they say 'Poor vehicle choice – a dirty and dangerous car', while the small print reminds guilty drivers that driving an inefficient 4x4 wastes the same amount of energy annually as leaving the fridge door open for seven years. See www.stopurban4x4s.org.uk.

On pages 90–1 there are some more comparative tables for different vehicles, sorted by CO_2 emissions per kilometre, with the best-performing at the top. Note the fuel-thirsty 4x4s and sports cars are right at the bottom of the class.

This is not an exhaustive list of every car model out there – if you are searching for a new or second-hand car but don't know its CO_2 emissions, you can easily find out by entering its vital statistics (manufacturer, model number, fuel type) into the database at http://www.vcacarfueldata.org.uk/ search/search.asp.

Calculate your car's carbon footprint

There are two major factors to consider when working out your car's carbon footprint. The first is what type of car you drive, if any. The second is how far you drive it. You can own the biggest Humvee on the block, but if you only drive it five miles per year, no one's going to complain.

On the other hand, you can own the most fuel-efficient Toyota Prius hybrid, but if you make a 50-mile-return commute each day to London, then your carbon budget is going to be pretty high.

So here's how to calculate your car's contribution to your carbon footprint.

Enter your car's emissions in grams per kilometre, using the table on pages 90–1 to find the correct figure or by searching for your model in the online database mentioned above.

Manufacturer	Model	Fuel type	Miles per gallon	Engine size	CO₂ emissions (g/km)
Toyota	Prius T Spirit Hybrid 1.5 VVT-i 5dr	Petrol	65.7	1.45	104
Ford	Fusion 1.6 Duratorq TDCi	Diesel	42.2	1.6	122
Smart	Fortwo coupé -Brabus 74 bhp	Petrol	53.3	0.74	127
Toyota	Yaris-1.0 VVT-i 3 & 5 dr	Petrol	52.3	1	127
Fiat	Punto 1.2 8V	Petrol	49.6	1.2	136
Peugeot	307 XSi 2.0 16v 5dr	Diesel	36.7	2	142
Renault	Mégane Coupé-Cabriolet	Diesel	51.4	1.9	154
Ford	Ka 1.3 Duratec	Petrol	45.6	1.3	154
SAAB	9-3 MY2006 5 dr Estate 1.9 TiDS (150 bhp)	Diesel	47.9	1.5	159
Vauxhall	Zafira, MY2006-1.9CDTi 120PS	Diesel	46.3	1.9	-165
Mazda	Mazda 6 – 2.0 TD 121 PS Estate	Diesel	42.8	2	172
Ford	Galaxy Zetec 1.9 TDI	Diesel	42.8	1.9	178
Jaguar	S Type – 2.7L Saloon	Diesel	41.5	2.7	179
Mercedes Benz	A Class - A200 Turbo 215 Tyres	Petrol	39.2	2	192

Volkswagen	Golf 2.0T GTI 5dr	Petrol	34.9	2	194
Hyundai	Tucson - 2.0i	Petrol	34.4	2	194
Kia	Sportage - 2.0	Petrol	34.4	2	194
Citroën	Xsara Picasso 2.0i 16v Desire	Petrol	32.9	2	205
Lotus	Elise	Petrol	38.2	1.8	208
Hyundai	Terracan - 2.9l CRTD 4WD	Diesel	32.4	2.9	241
Jaguar	S Type - 3.0L Saloon	Petrol	27.4	3	259
Audi	A3 Sport - 3.2 V6 quattro	Petrol	26.4	3.2	259
Land Rover	Discovery - 2.7 V6	Diesel	30	2.7	275
Mercedes Benz	M Class Estate	Petrol	19.3	5	319
BMW	X5 4.8is	Petrol	20.9	4.8	324
Cadillac	3.6 - V6 AUT	Petrol	24.4	3.5	349
Land Rover	Range Rover - 4.4 V8	Petrol	19	4.4	352
Mercedes Benz	Maybach	Petrol	15.9	5.5	383
Maserati	Coupe GT	Petrol	15.2	4.2	400
Bentley	Continental GT	Petrol	16.5	6	410
Ferrari	F430	Petrol	15.4	4.3	420
Lamborghini	Murcielago 147 Coupe	Petrol	13.2	6.1	500

Now multiply this figure by the number of kilometres you drive per year. (If you only know the mileage, take this figure and multiply it by 1.609 to convert to kilometres.)

If you don't know your distance travelled for the year, dig out your two most recent MOT certificates, each of which will have your total mileage on them. Subtract one from the other to get a year's worth of driving.

Number of kilometres driven per year

....

multiplied by the car's emissions (from above)

.... g/km.

Now divide by 1000 to get the total in kilograms.

If you are the only person in the car most of the time, then the carbon footprint is all yours. But if you are mostly sharing a car (perhaps there's two of you on a regular commute or the entire family drives to school and work together), you can divide the total emissions by the number of adults in the car.

> Total personal CO_2 per year from driving =
>
> .... kilograms

Keep this figure handy – we'll use it later to tot up your grand total.

Cutting your car carbs

Cutting your car-related emissions couldn't be easier. Just get rid of it and walk, cycle, or use public transport. Convinced? No? Well, thankfully, there are a few other options.

1. Share your car. In some busy areas, there are special lanes for cars with more than one occupant. Some stick-in-the-muds use a blow-up doll to pass muster, but if you want to halve your carbon footprint, you need a real person in the passenger seat. This will also reduce congestion and parking problems, and will mean that you can share costs.

2. Don't commute. Travelling twenty miles each way to work every day adds up to a scary 10,000 miles a year – with a heavy penalty in both emissions and frustration. These small daily drives soon add up to far more than the more occasional long-distance journeys. Consider working from home some days, reducing the need to travel. Many people find that they work more efficiently when outside the hubbub of the office. The rise in telecommuting offers significant potential to reduce transport-related emissions. In the longer term, plan to live near your office.

3. Get a model with the smallest possible engine size, cutting back on those CO_2 per kilometre totals from the table on pages 90–1. The government's car tax rates are now dependent on a vehicle's fuel efficiency, so you save money twice over by having a small car – once on tax, and secondly each time you go to the garage to fill up.

4. Don't use your car for short journeys, when emissions are at their worst. Try to use local shops and farmers' markets (on foot or by bike), which will be better for your community as well as good for the climate. Cycling is a healthy activity, keeping you fitter and happier.

5. Drive efficiently. Don't accelerate when there's a red light or a roundabout a hundred yards away. Cut your motorway top speed from 70mph to 60mph (most cars have an optimal cruising speed of 55mph in top gear) and change up gears before the engine revs rise too high. Remove roof-racks to minimize drag, and make sure your tyres are at the right pressure. Switch off the engine if you're stuck in traffic, and don't sit around revving the engine on cold mornings – just drive straight off. (Your neighbours will thank you for that too.) Stopping and starting in urban traffic uses much more fuel than cruising on the open road, though hybrids are better performers in urban areas because they can use electrical power for shorter distances.

Hybrids

In recent years, various models of hybrid cars have come onto the market. While initially very expensive, hybrids have begun to come down in price, and may even be cheaper once savings on fuel bills are taken into account – especially at a time of high oil prices.

Hybrids work by using petrol on the open road, but switching to electricity for short-burst city centre driving. Recharging happens automatically during braking, when some of the kinetic energy from the

---

Tip: One of the biggest causes of the rise in childhood obesity is the 'school run'. If your kids walk or cycle to school they will be burning up carbohydrate energy that might otherwise be stored as fat.

car's motion is converted to electrical energy and stored in a battery for future use.

Biofuels

Originally seen as the clean, green alternative, in recent studies biofuels have come in for some stick from environmentalists. Early biofuels were often made from recycled oil gathered from fish and chip shops, giving the exhaust of cars that used it a pleasant chippy smell, and recycling a resource that would otherwise be dumped. The benefits of biofuels are that no fossil fuel is used, so cars that burn it can go some way to being carbon-neutral.

Nowadays biofuels are much bigger business and most of the vegetable oil feedstock is imported. This can present problems, because carbon is used during transportation, and some vegetable oils – like palm oil – are often grown on land cleared of tropical forests, causing enormous releases of greenhouse gas pollution in the process. This so-called 'deforestation diesel' is worse than its conventional counterpart in terms of its impact on the climate. It is also wiping out the last remaining wild habitat of the orang-utan.

There are also concerns about burning a food resource – vegetable oil – in cars, rather than using it for, well, food. Growing crops to produce biofuels uses up valuable land space that could otherwise grow natural forests or food crops for people. It has been estimated that meeting the European Union's target (of 5% biofuels in its vehicle fleet by 2010) would require a quarter of Europe's agricultural land. If the petrol-thirsty US was to convert all its cars to run on biofuels, it would require four times the country's total arable area to grow the stuff. Cars would be happy, but humans would starve.

There have also been questions raised as to the energy return on biofuels. Some academics insist that it takes more energy to grow biofuels (because of all the oil used in tractors and processing) than they release on burning. There are many different views on the environmental benefits – or pitfalls – of biofuels, and until some consensus is achieved, it is safe to regard them with a healthy degree of scepticism.

Diesel

Many people are switching to diesel cars because per mile travelled they are much more efficient than petrol vehicles. By driving a diesel car you can save up to 10% of the carbon emissions that would come from sticking with petrol. On the other hand, diesel does have a fair few nasty pollutants associated with it, some of them – such as particulates – thought to be carcinogens and also associated with asthma and lung disease. These are hardly produced at all by petrol cars, which (if they have a catalytic converter in the exhaust) also produce fewer emissions of other toxins. There's no easy answer to this conundrum (other than not driving at all). However, given that global warming is all but certain, while the health effects of diesel emissions remain open to question, it is probably better to put the climate first.

Liquid petroleum gas (LPG)

LPG cars have lower CO_2 emissions than petrol-driven ones. They don't pay the London congestion charge for this reason. In carbon terms, LPG isn't any better than diesel, but it has much lower emissions of other pollutants (like particulates). There are currently around 1,300 LPG refuelling stations in this country.

Electric cars

Walk round the back of an electric car, and you can see that it produces no emissions at all – there isn't even an exhaust pipe! Problem solved? Not quite. It still uses energy – in this case electricity – which has to come from somewhere. Given that most electricity in the UK is still generated from fossil fuels (coal and gas), this simply shifts the emissions burden from the car exhaust pipe to the power station chimney. Because energy is lost in transmission over power lines, and because of inefficiencies in the generating process itself, it might actually have been less emissions-hungry to just burn petrol in the first place. That's unless you're charging your car from renewable sources only, in which case you get five stars and a pat on the back.

Alternatives to the car
Cycling
The more people who cycle, the better everyone's health and quality of life, with fewer cars jamming up the roads and more breathable air in cities. People often grumble about cyclists riding dangerously – through red lights and on pavements, for example – but such behaviour is frequently a reaction to the difficulties of sharing the road with far more dangerous cars and lorries. About 150 cyclists are killed each year in accidents (as compared to 1,500 car users – driving is still the most dangerous activity of all).

Tip: Think twice before ditching your car and buying a more efficient new one in order to save on fuel consumption. It takes a lot of energy to manufacture a car (10% of all the fuel used over the car's lifetime), so simply driving less or buying second-hand may be a better option.

Having said that, cyclists should obey the law just like other road users. If you live in the capital, the London Cycling Campaign is a great place to start – their vision is to make London a 'world-class cycling city'. See www.lcc.org.uk.

It goes without saying that the more bikes on the roads, the safer cyclists can feel. Motorists become more aware of pedal cyclists, and the political pressure for safer cycle routes increases. The campaigning organization Critical Mass organizes 'bike blockades' on a regular basis in different cities to promote cycling as an environmentally beneficial form of transport – unlike the car. See http://critical-mass.info.

Try to think of cycling as your main form of transport. It will increase your fitness, so perhaps you can save on those gym membership bills. There is nothing so absurd as seeing people driving to their local gym, going on the exercise bike for half an hour, and then driving home again. The problem with cars is that they allow us to travel great distances without actually moving. Nearly a quarter of UK adults are

---

Tip: Travelling by car is nine times more dangerous than travelling by train (2.8 deaths per billion passengers versus 0.3) and seven times more dangerous than by bus (0.4 per billion).

obese, whilst 27% of girls and 24% of boys aged 11 to 16 are also overweight. Reclaim the streets, and reclaim your health!

But can cycling really take over from the car? The answer is yes – more than a quarter of all journeys made by car are under two miles long, an easy cycling distance. The notorious 'school run' is an average journey distance of three miles, which can be cycled in less than half an hour, probably comparable to travelling by car when you factor in all the time spent sitting in jams.

The cycling promotion charity Sustrans points out that 45% of children say they would like to cycle to school, but only 4% are allowed to. In response, Sustrans has set up the Safe Routes to Schools and Bike It schemes – see the website www.saferoutesto schools.org.uk.

Sustrans has also set up a National Cycle Network of off-road and traffic-calmed cycle routes. The network now extends over 10,000 miles, and 75% of the

Tip: The CTC, a nationwide organization that campaigns on behalf of cyclists, may be able to provide legal assistance if you have an accident. See www.ctc.org.uk.

population lives within two miles of a route. It is well-signposted, and connects towns and villages, countryside and coast throughout the UK. See www.sustrans.org.uk. The website also has a useful map facility where you can enter your postcode and find the routes nearest to you.

Public transport

Buses and trains are both much safer than travelling by car, and their carbon emissions are dramatically lower too (see pages 105–10). It has never been easy to combine several different modes of transport (rail, bus, walking etc.) but the recently launched Transport Direct website has door-to-door information for the whole country. Just enter the postcode/village/town you are travelling to and from, and it will tell you how to do the journey – even with bus numbers, useful maps and journey times. See www.transportdirect.info.

MOTORBIKES

Motorbikes are more fuel efficient than cars on average, but then most of the time they only carry one person. The table opposite gives a comparison between some different models and their emissions, sorted by CO_2 emissions per kilometre. As with cars, your motorbike emissions will depend on the size of your vehicle's engine, and the distance you drive it

Manufacturer	Model	Miles per gallon	CO_2 emissions (g/km)
Yamaha	TW200-2005	75	87
Vespa	Granturismo 200-2005	61.9	105
Suzuki	C50T Boulevard	50	130
BMW	R1200ST-2005	48	136
Kawasaki	Z750S- 2006	45	144
KTM	625 SMC-2005	44.4	146
Harley-Davidson	FLSTNI Softail Deluxe- 2005	42	155
Buell	Lightning CityX XB9SX-2005	40.5	160
Honda	599-2005	40.6	160
Ducati	ST3- 2005	37.3	175
Victory	V92-2003	33.1	197

Motorbike fuel efficiency	CO_2 emissions in g/km
<35 mpg	217
35–45 mpg	161
>45 mpg	130

each year. Use the lower table on page 103 as a rough guide. If you want to be more precise, you can use the carbon calculator on www.carboncalculator.org. This is for cars, but if you select the mpg option then the emissions total will be the same.

So here's how to calculate your motorbike's contribution to your carbon footprint.

Enter your motorbike's estimated emissions in grams per kilometre, using the table on page 103, or the exact figure if you have it.

....

Multiply this figure by the number of kilometres you ride per year. (If you only have the mileage, multiply this figure by 1.609 to get the figure in km.) If you haven't kept a note of it, check your MOT certificates for the last two years, which will give the total mileage covered. Subtract one from the other to get last year's total.

Number of kilometres per year ....

multiplied by the bike's emissions (from above) =

.... grams.

Now divide by 1000 to get the total in kilograms.

Total CO_2 per year from motorbike riding =

.... kilograms

Keep this figure handy – we'll use it later to tot up your grand total.

PUBLIC TRANSPORT

Buses have bigger engines than cars, use up more fuel, and have higher emissions per mile travelled. But they also carry more people. At Britain's woefully low bus occupancy average of nine people per bus, they are no more efficient per person than a small car, but if you tend to use popular bus routes (such as city-centre routes in London, or rush hour in other places), then you're making big savings in terms of carbon.

Buses also have wider social gains over cars in that they don't use as much road space, so can carry a greater number of people short distances. Anyone who has seen city centres choked with car traffic, each car carrying just one person, knows how inefficient cars are as a way of getting large numbers of people around small spaces.

Trains are a much better environmental bet than cars, and they also have the advantage of making long-distance travel easy and safe (if not particularly cheap). With air travel being the worst of all possible options in carbon terms (see pages 110–13) the train is often the best alternative for inter-city and UK to continental travel. See Appendices A and B on pages 160–70 for some comparisons of the CO_2 emissions from different methods of travel.

Although trains are operated by different private companies, ticket sales and train information are combined through National Rail, and railcards for young people and senior citizens are also accepted across the network. For train times and ticket prices, contact National Rail Enquiries on 08457 484950 or www.nationalrail.co.uk.

Underground trains in London are electric-powered, so don't produce emissions directly into the atmosphere – luckily, as the tunnels would quickly get pretty smelly otherwise. But the electricity has to come from somewhere, so even travelling by underground has a carbon cost. London Mayor Ken Livingstone has floated the idea of getting the Tube running on renewable power, in which case you'll be able to skip this section. Meantime, see page 109 to work out London Underground travel's contribution to your carbon footprint. (For other urban rail systems outside London, like Newcastle's Metro or Glasgow's Underground, use the same multiplier.)

Ferries are seen by many as old-fashioned, and simply a way of getting one's car to the other side of the water. But travelling by ship – as well as being more pleasant than being squashed into a pressurized airline cabin – is also better for the climate. Ships come in all shapes and sizes, however: travelling to New York on the QEII is probably worse in carbon terms than going by air, because the ship is so large and luxurious. But the average cross-channel or North Sea ferry will get much better carbon mileage per passenger than an equivalent flight.

A safe rule of thumb is that the faster a ferry travels, the worse its emissions will be per passenger mile, because of the extra drag of the water. Super-fast catamarans may even be worse than flying, so be prepared to go slow.

Tip: Even if you give up your car and switch to public transport, regular miles travelled by commuting can soon stack up the carbon. Commuting 10,000 miles per year by train is the same as covering 3,000 miles in a car.

Calculate your annual emissions from public transport
It can be tricky to work out your total annual mileage
on public transport. Start by estimating the distance
travelled in an average week and multiply by 52.
Use Appendix A to find the distance covered in any
national journeys you have made.

Total number of miles travelled by bus:

....

Convert this to kilometres by multiplying by 1.609 and
then multiply by:
0.09 for London buses
0.17 for outside London
0.08 for express coach.

Total carbon cost from bus travel:

.... kilograms

Total number of miles travelled by train:

....

Convert this to kilometres by multiplying by 1.609 and
then multiply by:

0.11 for inter-city rail travel
0.16 for other services.

Total carbon cost from train travel:

.... kilograms

Total number of miles travelled by underground:

Convert this to kilometres by multiplying by 1.609
and then multiply by 0.07.

Total carbon cost from underground travel:

.... kilograms

Total number of miles travelled by boat:

....

multiply by 0.75 (or multiply the total distance in
kilometres by 0.47)

Total carbon cost from boat travel:

.... kilograms

The multiplier for boat travel is a very provisional figure, and depends very much on the type of boat you're travelling on (see above). You can try to do your own calculations by asking the crew what the ship's fuel consumption is and dividing it by the number of passengers.

Now add up your public transport figures:

Grand total of carbon emissions from public transport per year

.... kilograms

AVIATION

Aviation has seen explosive growth over the last few years, especially with the boom in budget airlines. Whereas travel to Spain or the Czech Republic was once out of bounds for price reasons, now stag parties can fly to Prague for a weekend at the cost of £20 per ticket – less than the stags are each likely to spend on beer in a couple of hours.

Plane travel has come in for an awful lot of flak recently, and not just from environmentalists. The Parliamentary Environmental Audit Committee, a cross-party group, in 2006 released a report arguing that it is 'scandalous' that airline fuel is untaxed, and that there is no VAT on air tickets. These keep airline tickets artificially cheap, and add up to a subsidy of billions of pounds each year to the industry.

While the cost to the passenger may be low, the cost to the climate is high. Britain's CO_2 emissions from aircraft doubled in the space of a decade, between 1990 and 2000, and are projected by the government to double again by 2030. According to a report by the Tyndall Centre, the UK would have to stop emissions from all other sectors of the economy in order to allow aviation to expand this dramatically and still meet our climate change targets.

Aviation's impact on the climate is worsened by the fact that the polluting emissions happen largely high up in the atmosphere, where they can do most damage. It's not only the old culprit CO_2 either – water vapour emissions (the contrails you see in the sky every day above Britain) also have a warming effect.

In total, the Intergovernmental Panel on Climate Change estimates the warming effect of aircraft emissions to be 2.7 times that of the carbon dioxide alone, because the emissions of other gases like nitrous oxides go into the upper atmosphere where they cause more damage. We'll factor this multiplier into our calculations below.

You will quickly see that a single long-haul return flight – say from London to Melbourne – will by itself double your whole year's carbon footprint. This isn't because aircraft are uniquely bad (the per-mile emissions are comparable to driving a car the same distance) but because the distances are so huge. Indeed, long-haul flights are more efficient per mile than short-haul ones, because extra fuel is used in take-off and landing, and makes up proportionately more of the total in shorter journeys.

The table on pages 114–15 gives some examples of CO_2 emissions from flights to common destinations, and you'll find more examples of short-haul national and European flights in Appendices A and B on pages 160–70.

For long-haul journeys, the best site on the web for calculating your flight emissions straightforwardly is http://chooseclimate.org/flying. You can simply click on your starting point and your destination on the interactive map, and it gives you your grand total in tonnes of CO_2 with the multiplier included.

Multiply by 1000 to get the total in kilograms.

If your geography isn't up to finding Lima and Milan, there are plenty of other easy-to-use carbon calculators on the web – two to choose from are www.carbonneutral.com/cncalculators/flight calculator.asp and www.carboncalculator.org.

Using different calculators may give you different estimates of the climate impacts of the same flight. This is because of different assumptions used. Some carbon offset companies don't use a multiplier at all. To be on the safe side, multiply the total given by the web calculators by 2.7. (If they give the total in tonnes, multiply by 1000 to get your CO_2 in kilograms.) Remember that the chooseclimate.org site already includes a multiplier, so no extra multiplication is needed if you use it.

> **Total carbon dioxide emissions from flights for the year**
>
> .... kilograms

Keep this figure handy – we need to add it to your total at the end.

Overseas holiday destinations	Distance in miles/km from London to capital city	CO_2 emissions in kg (including multiplier) for a one-way flight
Belgium	199mi/321km	163
France	213mi/343km	174
Netherlands	221mi/356km	181
Irish Republic	291mi/469km	239
Switzerland	465mi/748km	381
Germany	570mi/918km	468
Denmark	592mi/953km	486
Czech Republic	632mi/1018km	519
Austria	769mi/1237km	630
Spain	783mi/1261km	643
Italy	897mi/1444 km	736
Poland	904mi/1454km	741
Hungary	913mi/1469km	749
Portugal	985mi/1585km	808
Tunisia	1132mi/1821km	928

Bulgaria	1254mi/2017km	1028
Malta	1301mi/2093km	1067
Greece	1486 mi/2391 km	1219
Turkey	1761mi/2835km	1275
Cyprus	2000mi/3218km	1448
Egypt	2174mi/3499km	1574
United Arab Emirates	3407mi/5483km	2138
United States	3674mi/5913km	2306
Canada	3792mi/6102km	2379
India	4174mi/6718km	2620
Barbados	4211mi/6777km	2643
Jamaica	4686mi/7541km	2940
Sri Lanka	5413mi/8712	3397
Mexico	5556mi/8941km	3486
Thailand	5924mi/9534km	3718
South Africa	5968mi/9605km	3745
Australia	10554mi/16984km	5434

CONSUMPTION

We've done transport emissions and we've done household emissions – but add these two together and they comprise only two-thirds of the UK's carbon emissions. The missing third comes from the industrial sector: manufacturing, retail and services – emissions generated by companies producing products for consumers like you. Every item you buy in the shops, from strawberries to CDs, has an energy penalty associated with it – the energy in gathering the raw materials, in processing or manufacturing, transporting and then selling it to you.

You won't find this energy penalty on the label, of course, so you can't add it directly to your carbon total – and in any case, adding up everything in every cupboard would be beyond the call of duty for even the most ardent environmentalist. According to calculations by the Climate Outreach and Information Network, the average shopaholic will be adding three tonnes (3000kg) to their carbon budget each year just because of what they buy. In contrast, someone who finds their pleasures elsewhere, buying minimal

---

Tip: See http://coinet.org.uk/projects/challenge/measure for the Climate Outreach and Information Network's own carbon calculator, which can do the sums for you.

stuff (and second-hand where possible) may be adding as little as half a tonne. See Appendix C (pages 171–89) for some comparisons between big companies' approaches to climate change.

FOOD MILES

Perhaps the most important place to start is food. Have a quick rifle through your cupboards and fridge and jot down a note of the countries of origin for each food product.

The further the distance it has travelled, the bigger the carbon penalty. Each glass of orange juice, for example, 'contains' the equivalent of two glasses of petrol once the transport costs are included. Worse still are highly perishable fresh foods that have been flown in from far away – green beans from Kenya or lettuce from the US. They may be worth several times their weight in jet fuel once the transport costs are factored in. These are examples of 'food miles'.

A recent report suggested that a typical British Sunday roast, made with imported ingredients, would have seen its food constituents travel 81,000 km, the equivalent to two journeys around the Earth – with a total carbon cost of 3,700 grams of CO_2. On the other hand, had the beef, potatoes and vegetables all been

Product	Origin	Distance (km per kg)	CO_2 emissions (l/kg)	Fuel consumption
Strawberries, cherries, peppers, herbs	Ankara, Turkey	2835	1.61	0.64
Strawberries, green beans, salad onions, garden peas, sugar snaps	Cairo, Egypt	3520	2.0	0.8
Fine beans, mangetout, sugar snaps, courgettes, salad onion	Nairobi, Kenya	6804	3.87	1.54
Chanterelle mushrooms, various vegetables	Lusaka, Zambia	7905	4.5	1.79
Herbs	Harare, Zimbabwe	8257	4.7	1.87
Strawberries, baby spinach, raspberries, watercress, rocket, lettuce, limes, baby carrots,	Los Angeles, California	8774	5.0	1.98

cherry tomatoes, herbs, garlic, cranberries, cherries, broccoli, green beans			
Broccoli	Guatemala City, Guatemala	5.0	1.99
Avocados, cherry tomatoes, mangoes, limes	Mexico City, Mexico	5.1	2.02
Ginger	Rio de Janeiro, Brazil	5.3	2.11
Mangetout, fine beans, runner beans, herbs, asparagus	Bangkok, Thailand	5.4	2.16
Avocados, baby carrots, asparagus, radishes, peppers, grapes	Cape Town, South Africa	5.5	2.18
Garlic	Buenos Aires, Argentina	6.3	2.41
Raspberries, kiwi fruit, asparagus, grapes, plums, blackberries	Valparaiso, Chile	6.6	2.64
Beef cuts, ginger	Melbourne, Australia	9.6	3.83
Blueberries	Wellington, New Zealand	10.7	4.26

Source: 'Eating Oil', by Elm Farm Research Centre and Sustain, 2001. Contact: www.efrc.com. Reproduced with permission.

produced within a 48-km/30-mile radius, the carbon cost would have been a mere 58 grams of CO_2.

The table on pages 118–19 gives an indication of the emissions associated with different imported foods. Underline each type of produce that grows in, or in close proximity to, the UK. You'll quickly notice how many there are. So why aren't more British-origin vegetables and fruits in the shops?

Half of the answer is because of the vagaries of the global economy – cheaper labour overseas and relatively low fuel costs in transportation. The other half of the answer is that people have become accustomed to having all kinds of vegetables in all seasons. Mangetout and strawberries may grow in the UK in spring and summer, but they certainly don't in the winter. To buy local effectively, and to keep your food carbon budget down, you need to get used to cooking with in-season British ingredients, preferably grown as close to you as possible.

Opposite is a list of just some of the wide range of fruit and veg available in the UK in different seasons (including produce that can be stored, and which is grown in unheated cover like polytunnels). Plenty for everyone's taste, even in the dead of winter.

Spring	Summer	Autumn	Winter
New potatoes	New potatoes	Potatoes	Potatoes
Asparagus	Green beans	Apples	Apples
Strawberries	Broad beans	Pears	Pears
Spinach	Courgettes	Plums	Cabbage
Radish	Calabrese broccoli	Sprouting broccoli	Kale
Rocket	Tomatoes	Tomatoes	Turnips
Lettuce	Peppers	Lettuce	Carrots
Spring onions	Sweetcorn	Onions	Swede
Herbs	Mangetout	Wild mushrooms	Beetroot
Raspberries	Red onions	Garlic	Leeks
Green peas	Lettuce	Blackberries	Brussels sprouts
Rhubarb	Gooseberries	Cauliflower	Celeriac
Chinese cabbage	Summer squash	Pak choi	Winter squash
Leeks	Peppers	Celery	Pumpkin
Garlic	Aubergines	Pumpkin	Shallots
Pak choi	Cucumber	Marrow	Onions

Eating in season also gives you more cooking variety – instead of having the same meals on your household menu, week in, week out, you can vary what goes on the table with the changing seasons.

So, what is the best way to get hold of in-season local food? Some supermarkets, particularly the Co-op and Waitrose, are beginning to expand their 'buy local' lines. But supermarkets in general tend to choose homogenous fruit and veg that all looks the same (and tastes the same, too), thinking consumers will reject knobbly East Anglian carrots and misshapen Oxfordshire apples.

They also have very long supply lines – there are plenty of horror stories of supermarkets trucking vegetables from one end of the country to the other and then back again because of their centralized distribution systems. One supermarket even flew chives to Africa to be hand-tied and then back again, at a carbon cost that can only be guessed at!

However, the relentless rise in farmers' markets shows local food is coming back into vogue. From just a small base a decade ago, there are now hundreds of farmers' markets in towns and cities across the country, where you can get fresh produce straight from the producer

– cutting out the middleman altogether. To find your nearest one, visit www.farmersmarkets.net.

Organic food

Organic food isn't just healthier and tastier, it's generally better in carbon terms. One of the mantras of the organic movement is to 'buy local', eliminating excessive transport wherever possible. In addition, by taking the chemicals out of farming, one of the most energy-hungry agricultural processes is removed, further saving on emissions.

The best organic food will come from farmers' markets (see above) or can also be delivered to your door each week in an organic box scheme. Gone are the days when organic boxes just meant a few muddy turnips each Friday – nowadays fresh fruit and green veg are available to satisfy any taste. Organic cheese, milk, eggs and meats may also be available. There are literally hundreds of box schemes now around the UK – the Soil Association, the main organic certifying body, maintains an extensive list. See www.whyorganic.org.

Tip: Many box schemes include helpful recipes for the vegetables they deliver each week, so you don't stand gaping at celeriac or salsify wondering what to do with it!

Organic food isn't always the best choice, however. Many supermarkets have shelves bursting with organic produce – most of it imported. Over 70% of organic food in supermarkets is sourced from overseas. Even at the height of the British apple season, half the apples on sale at big supermarkets come from abroad – like Cox's all the way from New Zealand, for example. In this case, local trumps organic, as far as your carbon footprint is concerned.

Growing your own

You can reduce food miles almost to zero if you grow your own fruit and veg in your back garden. Keen gardeners can ensure a supply of fresh food almost year-round, with salads and sweetcorn in summer being replaced by potatoes, kale and parsnips in winter, in turn giving way to sprouting broccoli and broad beans in the spring.

If you don't have a garden, or you have one that only has space for a bit of lawn, then the best bet is to get a nearby allotment. Local councils have a duty to provide allotments for people who want them, and

Tip: Perhaps the best and most comprehensive grow-your-own book is the Henry Doubleday Research Association's *Encyclopedia of Organic Gardening*, available in paperback.

increasing numbers of people are taking on allotment plots – most of them organic. Rents are surprisingly low (as little as £20 a year for a decent-sized plot) and water is often provided on-site.

Most allotment sites have a contact number on a noticeboard at the entrance. Your local council will also be able to put you in contact with your nearest allotment association – typically you need to contact the leisure and parks department. Visit the National Society of Allotment and Leisure Gardeners for more info on www.nsalg.org.uk.

Waste

Another good reason for growing your own is so you can recycle kitchen waste through composting. Pretty much anything can go on the compost – don't believe that old wives' tale of not putting cooked food on it because it will attract rats. (Rats will probably feed from whatever you put on!)

Dump everything biodegradable on your compost heap, including non-recyclable paper and cardboard. Spice it up with a delicious regular dose of fresh urine to keep the microbes happy and within a few months you will have highly fertile compost to dig back into your garden or allotment.

Composting biodegradable waste is also important in preventing greenhouse gas emissions from landfill sites. Once dumped deep underground in a landfill, kitchen waste will generate methane – a greenhouse gas twenty times more powerful than carbon dioxide. Much better to rot it down in an aerobic environment like a heap at the bottom of your garden.

To make sure that your compost rots down properly without producing methane, turn it over every few weeks. Also, try to mix 'wet' waste (lawn mowings, kitchen scraps) with 'dry' waste (cardboard, straw) to keep the heap from getting too waterlogged. If it's getting wet and smelly, the chances are your compost heap is also producing methane! When it is rotting properly, it should be relatively odour-free and quite warm (even hot) inside.

As with other waste – the usual mantra of 'reduce, reuse, recycle' (in that order) applies. Buy less stuff to start with, and then reuse waste items such as jam jars for storing, well, jam. Most councils now collect recyclable materials like paper, glass and cans (and,

---

Tip: For more advice on composting, see the Centre for Alternative Technology's advice sheet: www.cat.org.uk/catpubs/display .tmpl?sku=ts_cs (costs 50p).

increasingly, plastic) in doorstep collections. Recycling materials saves energy because it avoids the need for new resources to be mined and drilled out of the ground, and recycling paper can help avoid deforestation – further protecting the climate.

But beware of long drives to a recycling centre – you can quickly negate any positive benefit from recycling if you have to use your car in the process. Some councils also send plastic waste to China for sorting, which involves a tremendous use of energy in shipping.

It is absolutely vital that you dispose of old fridges and freezers properly. Some older models contain ozone-destroying and climate-heating CFCs, and most coolants used today are still dangerous greenhouse gases and so need to be dealt with safely. Local councils have a responsibility to take your fridge – there may be free collection, or you could take it to a staffed recycling centre in your area. Contact your council's waste management department for more information.

Tip: It may not be easy to find out exactly what happens to the waste that your local council recycles. However, staff at the waste management department of your local authority should be able to give you some information.

WORKPLACE

Businesses are responsible for a large share of the UK's greenhouse gas emissions, and many companies haven't even begun to factor energy into their everyday decisions. See Appendix C (pages 171–89) for a report on the green credentials of some of the country's top companies.

How many offices do you pass by at night with all their lights blazing away, and their computers still humming? How many shops blast heat into the air outside by putting hot fans right in the doorway, while pouring cold out of open fridges into the aisle?

Unless you're the head honcho, your potential to limit your workplace's energy use might seem somewhat limited, but try to convince the powers that be that their bottom line could also be affected. Businesses that save energy also save money. The government's Climate Change Levy also penalizes private-sector operators for excessive energy use.

Here are some possible measures you could introduce (or lobby for) at work:

Tip: Contrary to what some techies claim, there is no evidence that turning computer monitors off does them any harm.

- Use natural ventilation (by opening windows) in summer, rather than air conditioning.
- Get the winter heat thermostat setting right, so it's not too hot and has everyone wandering around in shirtsleeves in mid-January.
- Get more fuel-efficient cars in the business fleet, and introduce incentives for car-pooling among staff. The Energy Saving Trust offers free fleet consultancy for businesses operating more than fifty vehicles. However, the same basic rules apply to everyone – see www.est.org.uk/fleet for more information.
- Introduce a policy of switching lights off in unused rooms, and turning computers off at the end of the working day.
- Get your workplace electricity supply switched to a renewable tariff, generated from green sources that don't contribute to CO_2 emissions (see pages 75–8).
- Reduce the amount of flying you do for work. Instead of seeing the overseas business trip as a prestigious perk, view it as a waste of valuable time, which is also highly damaging to the environment. (If it helps, visualize your expensive laptop being prised apart and ruined by spotty security officials at the airport.) Virtual meetings and teleconferencing may be a better way forward.
- Improve buying policies for shared workplace resources, with recycled paper for photocopying and

printing, and mains water at the cooler rather than big plastic bottles that need to be delivered regularly by truck. Make sure there are easy ways to recycle used paper, and introduce a policy of never printing out emails.

- Get an energy audit done of your company. The Carbon Trust offers free energy surveys – call 0800 085 2005 or visit www.carbontrust.co.uk for more information.

III
Sustainability

YOUR CARBON TOTAL

We've now looked at all aspects of your lifestyle, so it's time to come up with a definitive total of your carbon footprint. Flick back through all the separate sub-section totals and enter them here. Then add them all together and see how you do. The Great Green Hand is ready to judge you. Good luck.

Total carbon footprint from domestic gas use during the year

.... kilograms

Other domestic carbon use:

heating oil

.... kilograms

coal

.... kilograms

Total carbon footprint from domestic electricity use during the year

.... kilograms

Total carbon footprint from transport during the year

:...................... kilograms

We still need to factor in your lifestyle – what food you eat, how much stuff you buy. Given how difficult this is to quantify, use this shorthand:

- I have the latest of everything, love shopping, and eat mostly packaged convenience food: add 3000 kg
- I'm fairly thrifty, but buy new things when I need them and get most of my food from supermarkets: add 2000 kg
- I mostly grow my own organic food, shop locally, reuse and recycle, and wouldn't touch out-of-season strawberries with a bargepole: add only 600 kg

Now for the moment of truth....

Total personal carbon footprint from all sources

:...................... kilograms

So, how did you do? This tonnage figure doesn't mean much without some comparisons.

Average UK carbon footprint (kg CO_2 per person): 9,400 kg
Average US: 19,800 kg
Average China: 3,200 kg
Average India: 1,200 kg
Average Tanzania: 100 kg

This begs the question of what figure would be sustainable. According to the environmental writer Mayer Hillman, a fair rule of thumb is that each of us needs to get our personal carbon footprint down to about 1000 kg. That's a fairly daunting 90% cut from today's UK average, but is still achievable for anyone who follows the advice in this book to the letter.

Tip: The environmental writer and *Guardian* columnist George Monbiot suggests in his recent book *Heat: How to Stop the Planet Burning* that the UK can (and should) achieve across-the-board cuts in carbon emissions of 90% by 2030 without losing our comfort or security.

GENERATING YOUR OWN ENERGY

So far we have covered how you can cut back on energy use – but what about alternative renewable sources? How can you generate your own clean energy? For most of us, the easiest option is to sign up to a green electricity tariff (see pages 75–8), but why not help in the generating process? All you need, after all, is a bit of roof space, some sun and some wind.

SOLAR HOT WATER

For most people, this will be the most cost-effective solar option. In addition, since most houses get hot water from burning gas, this makes extra carbon savings even if you have already signed up to a green electricity tariff. Rock on!

Solar hot water options are proliferating but be sure that you contact a reputable company (see page 137) if you are looking at commercial installation options. The outfit that puts flyers through your door is not necessarily going to give you value for money. They might be more interested in getting their hands on the £400 government grant for which you may be eligible. (Note: there's a table explaining the grants available for help with energy-efficiency measure in the home on pages 142–43.)

So how good are solar thermal water systems? On bright days in summer you'll get a whole tankful of piping-hot water for the whole family. (Some people even use it for hot tubs or swimming pools, though frittering away scarce water and energy on such indulgences can hardly be recommended. Well, maybe just occasionally.) On grey winter days, you may get little or no heating effect, however.

What you'll need:

- Two to four square metres of full sun (no trees or shading from other roofs), southeast- to southwest-facing roof space, depending on how much water you're planning to heat.
- A compatible current hot water system: combi boilers, for example, are difficult and expensive to connect.
- Enough space to put in an additional water cylinder if required.
- Planning permission is necessary only if you live in a conservation area or listed building; otherwise, it's 'permitted development'.

Tip: The easiest hot water systems to convert to solar are ones that already have a cylinder in which they store the water that is heated by your boiler. If you have a combi boiler, contact the manufacturer to find out if it is compatible with solar hot water.

- A decent amount of cash: even with the grant, a system will set you back £2000–£4000.

What you'll get:
- Unlimited hot water on sunny days that doesn't cost a penny either in cash or carbon.
- On average, 60% of a household's hot water needs covered during the year.
- The chance to annoy the Joneses by being one step ahead of them in the solar energy revolution.

Contacts
To find out if your home is suitable, contact the government's Low Carbon Buildings Programme on www.lowcarbonbuildings.org. You can also call free on 0800 512 012. This is the organization that is also responsible for dishing out grants. These cover up to 30% of the cost of installation, up to a maximum of £400.

If you live in the Thames Valley area, a better starting point might be the Solarsavers project on

---

Tip: Be very cautious about which company you choose for a solar thermal installation. There are some cowboys about. For a list of government-accredited installers, see www.lowcarbonbuildings .org.uk/info/installers.

www.solarsavers.co.uk. This is a council-led initiative that liaises between householders interested in solar thermal systems and a small number of pre-selected reputable companies. They also have a free-phone number: 0800 107 0044.

If you live in London, you could be in luck. The Solar for London programme, launched by Mayor Ken Livingstone in 2003, provides grants of up to £1,500 on top of the standard government £400 on offer, depending on which borough you live in. See www.solarforlondon.org or call 0207 820 3156 for more information.

The DIY option

If you've got a combi boiler and very little cash, don't despair. Although you won't get a grant for a DIY scheme, you'll probably still save money (assuming you are – or are good mates with – a reasonably competent plumber).

All the components can be supplied as a kit, and at the simplest you can just put a separate 'solar' tap in your kitchen and bathroom. The Centre for

---

Tip: The average domestic solar hot water system reduces carbon dioxide emissions by a hefty 400 kg per year.

Alternative Technology has a publication called *Solar Water Heating: a DIY Guide*, priced at £5.99 (ISBN 1 898049 11 4).

You can buy an entire solar hot water system off the shelf (to install yourself) from the Low Impact Living Initiative, priced at a very reasonable £1,395. See www.lowimpact.org/acatalog/complete_LILI_solar_hot_water_kit.html.

SOLAR ELECTRICITY

So you've got the solar hot water, but there's still some south-facing roof (and money) left over. It's time to think about getting the sun to generate your electricity as well. Solar photovoltaic (PV) cells convert the sun's energy into electricity, which can then power lights and other appliances in your house.

Pros:
- You can sit inside on sunny days and watch your electricity meter run backwards as you sell surplus electricity back to the grid.
- You are saving approximately 325kg of carbon dioxide emissions per year, or eight tonnes over a PV system's average lifetime.
- By helping solar take off, you make it easier for others to follow your example.

- Generating your own power will make you even more aware of energy conservation and how much you use.
- There are many options for solar PV, including roof tiles that generate power, as well as the conventional flat panels for sloping roofs. You can even get clear panels that fit over conservatories.
- Because they have no moving parts, solar PV cells require very little maintenance, other than the odd wipe-down with a cloth.
- You can get grants under the government's low-carbon buildings programme of up to half the total cost, or about £3000 on a typical household installation.

Cons:

- You are unlikely to make a profit over the lifetime of the installation, given that solar PV costs about £6000–£7000 for a single house.
- You can't store electricity easily: battery banks are expensive and can be hazardous.
- Most domestic systems provide only about 30% of a household's typical use per year.

Tip: One green energy company has launched a home generation scheme, which will pay you in addition to the amount that you save on your regular bill. See www.good-energy.co.uk for more.

- Planning restrictions may apply in conservation areas or on listed buildings.
- If you are selling electricity back to the grid, the electricity company may only buy it back at wholesale price, rather than the higher retail price you pay. If this sounds like cheating, that's because it is.

Contact

There's information about the Low Carbon Buildings Programme on www.lowcarbonbuildings.org/myhome/generating/types/solarpv. If you live in Scotland, the relevant grant-giver is the rather long-winded Scottish Community and Householder Renewables Initiative. See www.est.org.uk/schri/ or telephone the SCHRI hotline on 0800 138 8858. It also provides grants for solar hot water and the other renewables options outlined below.

GRANTS

On pages 142–43 you'll find a table of grants available for domestic power generation, from the Low Carbon Buildings Programme, which covers the whole of the UK (except the Channel Islands and Isle of Man).

Tip: For advice on how to retro-fit your suburban 1930s semi with the latest in eco-innovation, see www.theyellowhouse.org.uk; the information is also available as a CD.

Technology	% Grant available (excl. VAT)
Solar photovoltaics	Maximum £3,000 per kW installed, up to a maximum of £15,000, subject to an overall 50% limit of the installed cost
Wind turbines	Maximum £1,000 per kW installed, up to a maximum of £5,000, subject to an overall 30% limit of the installed cost
Small hydro	Maximum £1,000 per kW installed, up to a maximum of £5,000, subject to an overall 30% limit of the installed cost
Solar thermal hot water	Maximum £400, regardless of size, subject to an overall 30% limit
Heat pumps Ground/water/air source	Maximum £1,200, regardless of size, subject to an overall 30% limit
Bio-energy 1. Room heater/stoves automated wood pellet feed	Maximum £600, regardless of size, subject to an overall 20% limit

2. Wood-fuelled boiler systems	Maximum £1,500, regardless of size, subject to an overall 30% limit
Renewable CHP (Combined heat and power)	Grant levels to be defined
MicroCHP (Combined heat and power)	Grant levels to be defined
Fuel cells	Grant levels to be defined

For more, see www.lowcarbonbuildings.org.uk

You will need to have completed the energy efficiency measures required by the programme before being entitled to a grant. You cannot double-fund a project if you live in Scotland or Northern Ireland – you can apply either to the Low Carbon Buildings Programme or to your regional agency. You can, however, add on funding from local authorities or other sources in addition to the government grant.

MICRO WIND TURBINES

If you live somewhere breezy, then a micro wind turbine may be the option for you. (Remember, these energy-generating measures aren't mutually

exclusive. You can have them all if you can afford it.)
We are often reminded that the UK is the windiest
country in Europe, and big wind turbines are
sprouting up at a rate of knots (if you'll excuse the pun).
Vitally important is that you check how windy your
exact location actually is, since the amount of power
generated goes up with the cube of the wind speed.
In plain English, this means that just small increases in
windiness mean that you get a lot more bang for your
buck. What? You don't have an anemometer? Better
check the online UK windspeed database in that case,
via www.bwea.com/noabl.

Siting is crucially important. If there are obstacles like
trees and other houses in the way of your prospective
turbine site, you may be wasting your time – they
both create turbulence (which can damage the
machine) and slow down the wind. That's why, to
be honest, you come across so few wind turbines
in towns.

Tip: If you generate more than you use at any one time, the extra
can be put back into the National Grid. There are two options here,
which depend on your electricity supplier. Some make a simple
calculation based on the size of your generating system and pay
you a monthly sum. Others will fit an export meter (at a cost of £30
to £100) to record how much electricity you supply to the grid.

Pros:

- Another source of clean, free power, which can cover an estimated 30% of your daily electricity use.
- The planning system is currently being simplified by the government to make it easier for households to install micro-wind generation.
- Much cheaper to buy and install than solar power: wind generators can cost as little as £1600, and are getting cheaper all the time. See www.windsave .com for a new to market domestic system. Another option is the Stealthgen, the turbine opted for by Conservative Party leader David Cameron. See www.stealthgen.com.
- Larger systems are an option in windier areas, costing between £4,000 and £18,000.
- Smaller systems can charge a battery, which can then be used to run household appliances. You can install them yourself from DIY kits costing as little as £650, available from www.windtrap.co.uk.
- Grants from the Low Carbon Buildings Programme run by the government will cover up to 30% of the costs.

Tip: Some new turbines are designed to be able to cope with turbulence, and can therefore be put on buildings rather than at the top of masts or towers (which might annoy the neighbours). One of these is the Swift, available from www.renewabledevices.com/

- Can reduce your household carbon dioxide emissions by 270kg or more per year.
- You make a visible, positive statement of your belief in a low-carbon future.

Cons:

- You are highly unlikely, unless you live in an extremely windy area, to be able to export a significant amount of unused power to the grid.
- Payback times may be very long, depending on the price you are paying for electricity.
- Possible noise and vibration, although most systems are engineered to avoid this as much as possible.
- You need an inverter to convert DC power from the wind turbine to the AC power of the mains (the same applies, however, to solar PV), which can cost £1000.

MICRO HYDRO

Only consider this option if you have a substantial stream or river running through your property – the garden pond won't do. Hydro-electricity is another unlimited renewable source: all you have to do is

Tip: The Centre for Alternative Technology produces an 'information sheet' on small-scale hydro, with links to books and other useful contacts. See www.cat.org.uk.

install a waterwheel and a generator and watch the cash roll in.

Well, that's the idea. On a practical level things are a bit more complicated. You need the water source to be fairly close to a grid connection (if you're planning to sell excess back to the grid) or to where you're going to use the power. It needs to be reliable, rather than drying to a trickle at the first sign of summer. And it needs to be flowing fairly rapidly to get the 'head' for substantial power.

Costs? We're probably talking £20–25K for a typical domestic scheme. As usual, grants are available. For more info, check out www.est.org.uk/myhome /generating/types/hydro.

GROUND SOURCE HEAT PUMP

Strange as it may seem, there is enough warmth in the ground to heat your house. All you have to do is capture and concentrate it with a heat pump – the same thing that your fridge does, except in reverse. It can even help to heat up water for showers and

---

Tip: Despite the hefty upfront costs, a ground-sourced heat pump may still be cheaper in the long-term than electric storage heaters or an oil-fired boiler, especially if you have a large property.

baths. One company in Oxfordshire heats its offices using a large ice cube in the basement. Don't ask us how it works, but it does.

What you need:

- Enough land for a trench 100 metres long or a spiral loop ten metres long; or a borehole where space is restricted.
- A 'ground loop': lengths of pipe laid in the trench that capture the ground's heat with flowing liquid.
- A heat pump, to take the heat from the ground loop and deliver it to heat your water for radiators or direct use.
- About £6,000–£9,000.
- A source of electricity, preferably green tariff from the grid or your own renewable supply.

What you get:

- Free heat all year round, derived from the sun, which is stored in the deep soil of your garden (although you still need to pay for the electricity to power it, which can be a substantial cost of several hundred pounds).
- Free cooling in hot weather if you have a system that can run in reverse.
- The pipe for the heat pump will be buried deep enough (about a metre) that it will not be affected by

frosts in a normal winter. The liquid inside also contains anti-freeze, just in case.

Contacts
Again, the Low Carbon Buildings Programme or its Scottish equivalent will provide grants if asked nicely. See www.est.org.uk/myhome/generating/types/groundsource/ and www.est.org.uk/schri/ (Scotland). The Ground Source Heat Pump Association also has more information and contacts for installers, via www.gshp.org.uk.

WOOD-BURNING STOVES

We mentioned earlier that your carbon score is effectively zero if you burn logs on a good wood-burning stove. (This is because trees are a renewable resource, where the carbon released through burning will be reabsorbed through regrowth.) While this is fairly low-end technology (in the sense that people have been doing it for hundreds of years), modern boilers can now heat your house (and water) without

---

Tip: Check that you are not in a smokeless zone before installing a wood burner. Under the Clean Air Act there are certain boilers you could still use, but check this with your local council. There are also building and safety regulations that need to be followed for any DIY installation.

you having to reload them every half hour. Wood pellets and wood chips, delivered into a large hopper for automatic feeding into the boiler, are a sustainable resource. (You'll need more space, though, than a simple living-room stove.) Boilers can also deliver you hot water, for long, satisfying baths on those dark winter evenings.

Contact

Grants are available, but only for wood boilers – the pretty cast-iron stove in the sitting room with the glass door you'll have to pay for yourself.

Make sure you source sustainable timber, by the way. Off-cuts from local tree surgeons are the best bet in urban and suburban settings. See www.est.org.uk/myhome/generating/types/biomass for more on types of appliances and available grants.

You can find your nearest logs or wood pellet supplier on the Log Pile website: www.nef.org.uk/logpile/fuelsuppliers/ index.htm.

Tip: Make sure that you buy a clean-burning stove to minimize the local pollution caused by burning wood, which can be dangerous to health.

TWO VIEWS OF THE FUTURE – SOLAR VS FOSSIL FUELS

Greenpeace view – solar power

The solar power industry is booming – more than 5000MW of power are now installed worldwide: equivalent to five big coal-fired power stations. In 2005 the European photovoltaics (PV) industry was worth more than five billion euros, and it continues to expand at a dramatic rate.

A report by Greenpeace and the European Photovoltaic Industry Association suggests that by 2025 solar PV systems could be generating about 589 TWh (terawatt hours) of electricity around the world. This would represent the annual output of 150 coal-fired power stations, and displace 353 million tonnes of carbon dioxide every single year.

The domestic-generating revolution will be central to this transformation: the report foresees that 290 million people worldwide would be generating their own electricity from grid-connected solar panels.

In developing countries, where there will be a huge growth in off-grid solar generation, a billion people could be using solar electricity.

However, even with such rapid growth, solar electricity would still only represent 3.5% of the global total in 2025. This could be expanded by 24% by 2040, establishing solar power as one of the largest sources of energy, and generating millions of new jobs in manufacturing and installation.

This won't happen by itself, however. In order for the solar revolution to gather speed, governments must introduce the following policies:

- support must be given to encourage consumers to install solar power
- subsidies must be removed from fossil fuels and nuclear power
- market mechanisms must be established to guarantee long-term high prices for solar electricity.

Conventional view – fossil fuels remain dominant

According to the International Energy Agency, which draws up the official projections for energy use in the future, by 2030 global demand for energy will have risen by more than 50% – much of it because of the rapid pace of industrialization in countries like India and China. Ominously, it also projects that 80% of this new growth will be met by fossil fuels – coal, oil and gas, with fast growth not just in the electricity sector but also in the demand for transportation fuels, where there is little prospect of a mainstream alternative to oil for decades to come. The result would be a catastrophic 52% rise in greenhouse gas emissions across the world, pushing us over the edge into rapid global warming (although the IEA stops short of drawing this conclusion).

While renewables would grow rapidly, their contribution would be swallowed up by the sheer growth in energy demand. The IEA projects that in a conventional business-as-usual scenario the share of renewable energy would grow from 8% in 2003 to a mere 9% in 2030.

Even under an optimistic 'Alternative Policy Scenario', where stronger efforts are made to promote renewable power, global CO_2 emissions would still rise by 30%, again because of the escalating demand for oil and gas in Asia, plus a rapid rise in world population.

Another potential fly in the IEA ointment is the issue of 'peak oil'. Many geologists are reaching the conclusion that supplies of oil are going to peak much sooner than the IEA projects, driving prices through the roof and forcing the world to switch to alternative fuel sources. In an optimistic scenario, this change would involve a more rapid substitution of renewables. However, a pessimistic (but equally realistic) scenario would see a switching back to the more plentiful but dirty coal as a power source – coal can be transformed into petrol through a chemical process, but even more greenhouse gas emissions would result.

So – which is it to be? Solar and renewables, or coal and gas? The answer is partly up to you.

DECENTRALIZED POWER – THE FUTURE?

Perhaps one day, rather than all-consuming centralized electricity from a national grid, many more of us will operate independently as small generators. Think internet versus old-style mainframes. The Micropower Council recently got a law passed by government promoting micro-generation, and environmental groups like Greenpeace are pushing hard for a small-

---

Tip: Are we headed for the era of the hydrogen economy? Jeremy Rifkin, president of the Foundation on Economic Trends, thinks so. In his book, *The Hydrogen Economy*, Rifkin outlines what he calls the 'worldwide hydrogen energy web', where most energy users would also be energy generators, and all of it based on renewable power.

generating future – particularly as an alternative to big generating plants like nuclear and coal stations.

One of the worst things about current fossil fuel-powered power stations is the fact that they waste over 60% of the heat generated by burning coal or gas. You can see their cooling towers pouring this waste heat into the atmosphere. So much heat is wasted that it could heat every building in the UK if used in a 'combined heat and power' system like those they have in Sweden and Denmark.

Now the suggestion is for combined heat and power (CHP) on a much smaller scale – at the level of the individual household. Whether fired by gas or biomass, or perhaps even fuel-cells, each home could be generating electricity and using the heat produced

---

Tip: The first domestic CHP system is now available in the UK, called the WhisperGen. Similar in size and shape to a domestic dishwasher, it is supplied by the electricity company Powergen and runs off your gas supply, generating heat for your home at the same time as electricity for your TV and stereo. According to Powergen, it could save 1,500kg of carbon dioxide emissions per year for the average household. It could also save £150 a year on your energy bills, and is priced at £3000 including installation. Contact: 0800 068 6515.

simultaneously. It sounds ambitious, but micro-CHP is already becoming available, and is well worth considering if you want to replace an old boiler anyway and are heating a large property or space. Upfront it may be slightly more expensive, but you'll save both money and carbon in the long run.

VISIONS OF A SUSTAINABLE BRITAIN

Hopefully by now this book has convinced you that living in a sustainable society won't necessarily involve woolly jumpers, cold water, sandals and beards. We have the tools in our grasp not just to make the jump to a low-carbon future, but also to improve our quality of life hugely. Here's how the country might change as everyone makes the transition to a low-carbon economy:

- Local shops make a comeback in cities, towns and villages.
- Communities see a renaissance as people meet more in the streets rather than going everywhere by car.
- Farmers get better prices for their products, selling direct to consumers rather than being beaten down by powerful supermarkets.
- Consumers get better, tastier food, with less need for the packaging and preservation that reduces nutritional value and quality.

- Less fuel poverty, which particularly afflicts the old, as better insulation and energy conservation makes houses warmer and more comfortable for less money.
- Children can play outside without constant fear of being run over, instead of living in virtual captivity in front of the TV or computer.
- The proliferation of micro-generation technologies puts people in control of their own energy supplies, instead of leaving them dependent on the vagaries of overseas markets (and despots in oil-producing countries).
- Health indicators will improve as people are more active, with less obesity, heart disease and other afflictions of over-consumption.
- People understand more about where their energy comes from and why it is important to use it wisely and sustainably.
- A reduction in conflicts over scarce fossil fuel supplies, particularly oil from unstable or repressive regimes.
- A quieter, slower life, where people take more time when travelling and travel less.
- We all learn to ride horses again – surely the ultimate in rural green transport.
- They close the motorways and airports down, and convert them back into forests.

Okay, maybe we're getting a bit carried away there. But the message should be clear. Transforming our lifestyles for a low-carbon future won't necessarily be a nasty sacrifice. It could in fact be a life-enhancing boon for everyone except the most committed of petrol-heads. But even more important in the longer term, by becoming a low-carbon consumer you will be helping to bring down global greenhouse gas emissions.

While the sad truth is that it may already be too late to save some of the most vulnerable species and ecosystems (because of global warming already in the pipeline), if global emissions can be reduced soon we will have a good chance of avoiding the more catastrophic impacts of climate change.

By doing what you can, with the aid of this book, you will be taking responsibility for helping to ensure that our children and grandchildren can continue to live on a habitable planet.

Appendices

APPENDIX A:					
CO$_2$ emissions for journeys within the UK					
Journey	no. of miles	no. of km	by car	by train	by plane
From Aberdeen					
to Birmingham	431	693	134	76	353
to Bristol	513	825	160	91	420
to Cardiff	531	854	166	94	435
to Carlisle	231	372	72	41	–
to Edinburgh	125	201	39	22	–
to Glasgow	147	236	46	26	–
to Liverpool	357	574	111	63	293
to London	544	875	170	96	446
to Manchester	352	566	110	62	289
to Newcastle	234	376	73	41	192
to Norwich	488	785	152	86	400
to Plymouth	628	1010	196	111	515
to Southampton	570	917	178	101	468
to York	322	518	100	57	264
From Birmingham					
to Aberdeen	431	693	134	76	353
to Bristol	90	145	28	16	–
to Cardiff	109	175	34	19	–
to Carlisle	199	320	62	35	–
to Edinburgh	298	479	93	53	244
to Glasgow	296	476	92	52	243
to Liverpool	102	164	32	18	–
to London	120	193	37	21	–

to Manchester	89	143	28	16	–
to Newcastle	208	335	65	37	–
to Norwich	159	256	50	28	–
to Plymouth	205	330	64	36	–
to Southampton	135	217	42	24	–
to York	134	216	42	24	–

From Bristol

to Aberdeen	513	825	160	91	420
to Birmingham	90	145	28	16	–
to Cardiff	44	71	14	8	–
to Carlisle	282	454	88	50	232
to Edinburgh	380	611	119	67	311
to Glasgow	379	607	118	67	310
to Liverpool	184	296	57	33	–
to London	120	193	37	21	–
to Manchester	171	275	53	30	–
to Newcastle	299	481	93	53	245
to Norwich	233	375	73	41	191
to Plymouth	125	201	39	22	–
to Southampton	106	170	33	19	–
to York	225	362	70	40	184

From Cardiff

to Aberdeen	531	854	166	94	435
to Birmingham	109	175	34	19	–
to Bristol	44	71	14	8	–
to Carlisle	300	483	94	53	246
to Edinburgh	398	640	124	70	326

Journey	no. of miles	no. of km	by car	by train	by plane
From Cardiff, contd					
to Glasgow	397	639	124	70	326
to Liverpool	202	325	63	36	–
to London	152	245	48	27	–
to Manchester	189	304	60	33	–
to Newcastle	317	510	99	56	260
to Norwich	264	425	82	47	217
to Plymouth	152	245	48	27	–
to Southampton	138	222	43	24	–
to York	243	391	76	43	199
From Carlisle					
to Aberdeen	231	372	72	41	–
to Birmingham	199	320	62	35	–
to Bristol	282	454	88	50	232
to Cardiff	300	483	94	53	246
to Edinburgh	98	158	31	17	–
to Glasgow	96	154	30	17	–
to Liverpool	126	203	39	22	–
to London	313	504	98	55	257
to Manchester	121	195	38	21	–
to Newcastle	60	96	19	10	–
to Norwich	282	453	88	50	144
to Plymouth	396	637	124	70	325
to Southampton	339	545	106	60	278
to York	117	188	36	21	–

From Edinburgh

to Aberdeen	125	201	39	22	–
to Birmingham	298	479	93	53	244
to Bristol	380	611	119	67	311
to Cardiff	398	640	124	70	326
to Carlisle	98	158	31	17	–
to Glasgow	49	79	15	9	–
to Liverpool	226	364	71	40	186
to London	413	665	129	73	339
to Manchester	221	356	69	39	182
to Newcastle	106	171	33	19	–
to Norwich	359	578	112	64	294
to Plymouth	496	798	155	88	407
to Southampton	439	706	137	78	360
to York	193	311	60	34	–

From Glasgow

to Aberdeen	147	236	46	26	–
to Birmingham	296	476	92	52	243
to Bristol	379	607	118	67	310
to Cardiff	397	639	124	70	326
to Carlisle	96	154	30	17	–
to Edinburgh	49	79	15	9	–
to Liverpool	222	357	69	39	182
to London	409	658	128	72	336
to Manchester	217	349	68	38	178
to Newcastle	154	248	48	27	–
to Norwich	379	610	118	67	311
to Plymouth	492	792	154	87	404

Journey	no. of miles	no. of km	by car	by train	by plane
From Glasgow, contd					
to Southampton	435	700	136	77	357
to York	213	343	67	37	175
From Liverpool					
to Aberdeen	357	574	111	63	293
to Birmingham	102	164	32	18	–
to Bristol	184	296	57	33	–
to Cardiff	202	325	63	36	–
to Carlisle	126	203	39	22	–
to Edinburgh	226	364	71	40	186
to Glasgow	222	357	69	39	182
to London	215	346	67	38	176
to Manchester	35	56	11	6	–
to Newcastle	176	283	55	31	–
to Norwich	241	388	75	43	198
to Plymouth	298	479	93	53	244
to Southampton	241	388	75	43	198
to York	103	166	32	18	–
From London					
to Aberdeen	544	875	170	96	446
to Birmingham	120	193	37	21	–
to Bristol	120	193	37	21	–
to Cardiff	152	245	48	27	–
to Carlisle	313	504	98	55	257
to Edinburgh	413	665	129	73	339

to Glasgow	409	658	128	72	336
to Liverpool	215	346	67	38	176
to Manchester	202	325	63	36	166
to Newcastle	284	457	89	50	233
to Norwich	115	185	36	20	–
to Plymouth	241	388	75	43	198
to Southampton	80	128	25	14	–
to York	211	339	66	37	173

From Manchester

to Aberdeen	352	566	110	62	289
to Birmingham	89	143	28	16	–
to Bristol	171	275	53	30	–
to Cardiff	189	304	60	33	–
to Carlisle	121	195	38	21	–
to Edinburgh	221	356	69	39	182
to Glasgow	217	349	68	38	178
to Liverpool	35	56	11	6	–
to London	202	325	63	36	166
to Newcastle	146	235	46	26	–
to Norwich	186	299	58	33	–
to Plymouth	286	460	89	51	146
to Southampton	228	367	71	40	187
to York	72	116	22	13	–

From Newcastle

to Aberdeen	234	376	73	41	192
to Birmingham	208	335	65	37	–
to Bristol	299	481	93	53	245

Journey	no. of miles	no. of km	by car	by train	by plane
From Newcastle, contd					
to Cardiff	317	510	99	56	260
to Carlisle	60	96	19	10	–
to Edinburgh	106	171	33	19	–
to Glasgow	154	248	48	27	–
to Liverpool	176	283	55	31	–
to London	284	457	89	50	233
to Manchester	146	235	46	26	–
to Norwich	254	409	79	45	208
to Plymouth	413	664	129	73	339
to Southampton	324	521	101	57	266
to York	89	143	28	16	–
From Norwich					
to Aberdeen	488	785	152	86	400
to Birmingham	159	256	50	28	–
to Bristol	233	375	73	41	191
to Cardiff	264	425	82	47	217
to Carlisle	282	453	88	50	144
to Edinburgh	359	578	112	64	294
to Glasgow	379	610	118	67	311
to Liverpool	241	388	75	43	198
to London	115	185	36	20	–
to Manchester	186	299	58	33	–
to Newcastle	254	409	79	45	208
to Plymouth	354	570	110	63	180

to Southampton	193	310	60	34	–
to York	180	191	37	21	–

From Plymouth

to Aberdeen	628	1010	196	111	515
to Birmingham	205	330	64	36	–
to Bristol	125	201	39	22	–
to Cardiff	152	245	48	27	–
to Carlisle	396	637	124	70	325
to Edinburgh	496	798	155	88	407
to Glasgow	492	792	154	87	404
to Liverpool	298	479	93	53	244
to London	241	388	75	43	198
to Manchester	286	460	89	51	146
to Newcastle	413	664	129	73	339
to Norwich	354	570	110	63	180
to Southampton	152	244	47	27	–
to York	340	547	106	60	173

From Southampton

to Aberdeen	570	917	178	101	468
to Birmingham	135	217	42	24	–
to Bristol	106	170	33	19	–
to Cardiff	138	222	43	24	–
to Carlisle	339	545	106	60	278
to Edinburgh	439	706	137	78	360
to Glasgow	435	700	136	77	357
to Liverpool	241	388	75	43	198

Journey	no. of miles	no. of km	by car	by train	by plane
From Southampton, contd					
to London	80	128	25	14	–
to Manchester	228	367	71	40	187
to Newcastle	324	521	101	57	266
to Norwich	193	310	60	34	–
to Plymouth	152	244	47	27	–
to York	251	404	78	44	206
From York					
to Aberdeen	322	518	100	57	264
to Birmingham	90	145	28	16	–
to Bristol	225	362	70	40	184
to Cardiff	243	391	76	43	199
to Carlisle	117	188	36	21	–
to Edinburgh	193	311	60	34	–
to Glasgow	213	343	67	37	175
to Liverpool	103	166	32	18	–
to London	211	339	66	37	173
to Manchester	72	116	22	13	–
to Newcastle	89	143	28	16	–
to Norwich	180	191	37	21	–
to Plymouth	340	547	106	60	173
to Southampton	251	404	78	44	206

APPENDIX B:

CO_2 emissions for return journeys to Europe

Journey	no. of miles	no. of km	by plane	by train	by ferry
Austria (London to Vienna)					
	1538	2474	1260	272	–
Belgium (London to Ostend)					
	246	396	188	44	186
Channel Islands (Southampton to Jersey)					
	252	406	207	–	191
Czech Republic (London to Prague)					
	1266	2037	1038	224	–
Denmark (Newcastle to Esbjerg)					
	804	1294	659	–	608
France (London to Paris)					
	426	686	348	75	–
(Portsmouth to St Malo)					
	324	522	266	–	245
Germany (Newcastle to Hamburg)					
	956	1538	784	169	723
(London to Berlin)					
	1140	1836	936	202	–
Greece (London to Athens)					
	2972	4782	2438	526	–

Journey	no. of miles	no. of km	by plane	by train	by ferry
Ireland (Liverpool to Dublin)					
	280	450	230	–	212
Italy (London to Rome)					
	1794	2888	1472	159	–
Netherlands (Newcastle to Amsterdam)					
	652	1050	535	–	494
Northern Ireland (Glasgow to Belfast)					
	222	358	182	–	168
Norway (Newcastle to Bergen)					
	887	1430	729	–	672
Spain (Portsmouth to Bilbao)					
	1056	1700	867	–	799
(London to Madrid)					
	1566	2522	1286	277	–
Sweden (Newcastle to Gothenburg)					
	1109	1784	910	–	838
Switzerland (London to Bern)					
	930	1496	762	164	–

APPENDIX C:

Corporate carbon

Things are going to get a little complicated if you try to tot up the carbon used by your high-street bank each time you go to deposit a cheque, for example. Every company, from the smallest corner shop to the biggest mega-corporation, uses energy and has associated greenhouse gas emissions. And you the consumer have a choice about which companies to support. So how are we to rank them?

Each year an organization called the Carbon Disclosure Project sends a questionnaire to the 500 largest companies in the world (the so-called FT500) asking them what they are doing about their own greenhouse gas emissions. The request for information was backed in 2005 by 155 signatures of some of the largest global investors, representing $21 trillion (£11.3 trillion) in assets.

Here are the responses from some of the companies contacted.

Company	Greenhouse gas emissions (tonnes)	Emissions reduction programme?	Sample comments
AstraZeneca (pharmaceuticals)	294,000	20% reduction by 2005 (achieved)	'We believe that AZ has a relatively small impact on climate change for such a large organization, but are working hard to make sure that that impact is minimized.'
Aviva UK (insurance; owns RAC and Norwich Union)	70,156	75% in UK achieved	'Aviva has identified climate change as one of the top ten risks which faces our company.'
Avon US (women's beauty products)	Information still being compiled	Energy efficient office lightbulbs	'Although our industry sector is not energy intensive, we continue to focus on reducing energy consumption and set goals accordingly.'

Barclays Bank	182,117	10% in buildings, 6% in business travel	'During the past 12 months the Environmental Services Team has financed the development of a number of renewable energy projects including onshore wind, landfill gas, small scale hydro and biomass plants.'
BG Group UK (British Gas)	4,162,886	Various at local level	'Our main contribution to helping to tackle climate change comes through our core business, natural gas, the lowest carbon fossil fuel.'
BP UK (oil/energy)	63,200,000 (1,376,000,000 including oil produced by BP)	10% reduction by 2010 achieved in 2001	'The prospect of climate change is a challenge that we must address if we are to fulfill our aspiration to be a sustainable company in a sustainable world.'

Company	Greenhouse gas emissions (tonnes)	Emissions reduction programme?	Sample comments
British American Tobacco	1,167,865	5.2% reduction by 2008 achieved in 2004	'We strive to continuously reduce the carbon dioxide emissions from our operations.'
British Sky Broadcasting	36,491	10% reductions by 2010	'Sky achieves emissions reductions through the use of building systems and design such as free cooling, chilled beams, efficient chillers, efficient lighting, presence detection and modular boilers.'
BT UK (telecoms)	822,697	71% reduction achieved, 25% reduction by 2010 is target	'Climate change has the potential to affect BT's operations since the associated extreme weather conditions

Cadbury Schweppes UK (food and drink)	1,101,400	1% reduction in energy per tonne of product per year	such as gales and floods can lead to network disruption.'
Centrica UK (energy)	12,834,333	First wind power coming onstream in 2005	'Cadbury Schweppes believes that Climate Change is one of the greatest environmental challenges facing mankind.'
Chevron-Texaco (oil)	63,500,000	Various schemes on energy efficiency	'Centrica owns six gas-fired power stations [and] has therefore developed a strategy to mitigate the impact of emissions trading on our operation.'
			'Chevron recognizes and shares the concerns that governments and the public have about climate change.'

Company	Greenhouse gas emissions (tonnes)	Emissions reduction programme?	Sample comments
Cisco Systems (computers)	In preparation	10% electricity renewably sourced	'We are continuously reviewing opportunities to improve the energy efficiency of our products.'
Coca-Cola Company	5,700,000	Moving away from refrigerants that are greenhouse gases	'The Coca-Cola Company takes the issue of climate change very seriously and we are working to reduce greenhouse gases so that we can continue to operate in an environmentally responsible manner.'
Colgate (toothpaste)	737,000	15% reduction in energy per tonne of product by 2006	'While Colgate is not an energy-intensive company, we are working to continuously

Daimler-Chrysler	7,050,000	Further reductions in car fuel consumption	improve our energy efficiency around the world.' 'We are consistently working to further reduce carbon-dioxide emissions both at our production sites and with our products.'
Diageo (drinks: brands include Guinness, Smirnoff, Baileys)	784,000	Reductions in emissions per litre to 241g (from 255) by 2007	'The emission of greenhouse gases is seen by Diageo as one of its main environmental impacts.'
Du Pont (chemicals)	25,000,000	Emissions reduction of 72% achieved	'We began taking action to reduce greenhouse gas emissions in the early 1990s, have accomplished major global reductions and set ambitious goals for the current decade.'

Company	Greenhouse gas emissions (tonnes)	Emissions reduction programme?	Sample comments
Exxon-Mobil (Esso)	138,000,000 from own operations	Reduction in flaring, efficiency improvements in refineries	'ExxonMobil recognizes that the potential impacts of greenhouse gas emissions on society and ecosystems may prove to be significant, although scientific evidence remains inconclusive.'
Ford Motor Company	8,400,000	5% reduction target within the UK	'We are working to reduce greenhouse gas emissions proactively, affordably and in line with the interests of our share-holders and other stakeholders.'
Gap (clothing)	753,174	See right	'We continue to identify, evaluate and develop strategies

Gillette (men's products)	496,494	Emissions reduced 14% since 2000	'to implement energy conservation opportunities at our owned and operated facilities with an emphasis on our Gap brand stores.'
			'Gillette has long-standing commitments to natural resource conservation and environmental protection.'
Glaxo-SmithKline (pharmaceuticals)	2,082,000 (CO_2 only)	8% per unit sales by 2005	'In 2004 at Barnard Castle in the UK two wind turbines have been installed and can provide up to 10% of the site's electricity when running at full speed.'
Halliburton	5,511,000	See right	'Halliburton has some limited GHG emission reduction programmes in place. These are

Company	Greenhouse gas emissions (tonnes)	Emissions reduction programme?	Sample comments
Halliburton, contd			generally local energy reduction programmes initiated by local operations personnel. There is no corporate-wide emissions reduction programme in place.'
HBOS (Bank of Scotland etc)	181,023	Buys renewable power in UK; 5% reduction over next five years	'HBOS is working with WWF-UK to support the development of its One Million Sustainable Homes Campaign.'
Heinz	1,004,000	Replacing boilers in Wigan facility	'For the time being, Heinz has opted out of the EU Emissions Trading Scheme until January 2007.'

HSBC (banking)	585,000	Have reduced UK emissions by 33%	'[Climate change] is a major issue for our customers and our staff, as well as for every organization on the planet, no matter how large or how small.'
IBM (computers)	2,194,400 CO_2 and 228,000 PFC	4% reduction per year over six years up to 2005 exceeded	'IBM realizes that such climate change could impact the economy and the quality of life for current and future generations.'
Johnson & Johnson (healthcare)	942,900	See right	'In 1999, in partnership with the World Wildlife Fund, Johnson & Johnson announced a worldwide corporate goal to reduce CO_2 emissions 7% below 1990 levels by 2010.'

Company	Greenhouse gas emissions (tonnes)	Emissions reduction programme?	Sample comments
Kingfisher (owns B&Q)	Data being collected	Reduction of 50% emissions per cubic metre of product by 2005 achieved	'For the past six years all new stores have incorporated a building energy management system. Planning objections continue to inhibit our installation of wind turbines although we continue to actively investigate opportunities.'
Kraft Foods	3,040,000	Emissions stable despite 10% increase in production	'We are focused on overall energy reduction and more specifically, efficient use of fossil fuels and electricity in our manufacturing processes where

	equipment such as boilers, baking ovens and coffee roasters are utilized.'		
L'Oreal (hair and skincare)	87,248	Switching to natural gas	'Trains and ships are integrated elements in our transport system.'
Legal & General UK (insurance)	30,578	See right. Also videoconferencing and company car efficiencies	'Our three largest properties have achieved a 15% reduction in energy consumption during the period 2001–04.'
Marks & Spencer	372,000	10% renewable energy contracts	'Remote monitoring of store refrigeration and heating, ventilation and air conditioning systems have saved around 7,000 tonnes of CO_2 p.a. The initiative won the Energy Efficiency category of the 2005

Company	Greenhouse gas emissions (tonnes)	Emissions reduction programme?	Sample comments
Marks & Spencer, contd			*Sunday Telegraph*/ Carbon Trust Innovations Awards.'
Microsoft	260,510	Subsidizes public transport for employees, energy reductions in new buildings.	'Microsoft has a programme to build energy efficiency functionality into our next generation of Microsoft Office software.'
Motorola	196,000	25% reduction in energy use (exceeded – see right)	'To date, we have made the following progress: 78% reduction in energy use from 2000 to 2004.'
National Grid	10,595,000	Replacing gas mains	'Methane emissions arising from leakage and venting

Nestlé	4,410,000	Various	account for some 50% of our total GHG emissions.' 'At every step along the supply chain, Nestlé addresses environmental aspects.'
Nike	110,610	13% reduction by 2005 from 1998	'Nike will continue to demonstrate its support for the objective of the Kyoto Treaty – reducing human created emissions that contribute to climate change.'
Proctor and Gamble (cleaning, nappies etc)	2,937,000	See right	'Our energy eco-efficiency is more than 71% greater than in 1990, meaning there is 71% more product being made per unit of energy.'

Company	Greenhouse gas emissions (tonnes)	Emissions reduction programme?	Sample comments
Royal Bank of Scotland	387,000	Reduce emissions by 5% by 2005	'In 2004, the RBS group and the Carbon Trust announced an energy efficiency partnership to help the RBS further reduce energy consumption and CO_2 emissions at our sites.'
Royal Dutch/Shell (oil)	112,000,000	5% annual reduction in emissions	'We are one of the world's largest manufacturers of solar cells. Over twenty-five years Shell Solar has supplied solar cells and modules with a peak power of over 350 megawatts. This represents around one fifth of total solar power installed worldwide.'

Scottish Power	56,947,802	Renewables investment	'To meet the growing demand, we have undertaken significant investments in gas-fired generation and renewables.'
Sony	2,088,485	See right	'For greenhouse gas emissions, Sony targets to raise eco-efficiency by 1.5 times by fiscal 2005 and 2 times by fiscal 2010, compared with fiscal 2000. Eco-efficiency of Greenhouse Gas Index is calculated by Sales divided by Environmental Impact.'
Starbucks	254,000	See right	'Our first action has been to purchase enough wind energy, using renewable energy certificates, to match 5% of the

Company	Greenhouse gas emissions (tonnes)	Emissions reduction programme?	Sample comments
Starbucks, contd			energy needed to power Starbucks Company-operated stores in North America in fiscal 2005.'
Toyota	5,840,000	Currently 19% below 1990 level	'Toyota has made environmental responses a top management priority and has been taking actions accordingly, as exemplified by our work toward achieving zero emissions.'
Unilever (food and cleaning products)	3,600,000	30% reduction achieved 1995-2004	'Climate change is one of the most pressing and complex global environmental issues.'

| Vodafone | 1,300,000 | Plans in progress | 'Climate change has been prioritised by Vodafone as an environmental issue for particular attention.' |
| Volvo | 293,500 | More fuel-efficient vehicles | 'Volvo Group has a future fuels strategy and has also adopted very far-reaching objectives for the Group on energy savings both in our industrial activities and in our products.' |

USEFUL ADDRESSES

Energy Saving Trust
www.est.org.uk
0800 512012
Government-funded energy
advisory body for the
general public

Carbon Trust
www.carbontrust.co.uk
0800 0852005
Government-funded energy
advisory body for business

**Climate Outreach and
Information Network**
www.coinet.org.uk
01865 727911
Education and resources on
climate change and energy

**Centre for Alternative
Technology**
www.cat.org.uk
01654 705950
Machynlleth-based green
technology centre, with
excellent visitors' centre

Climate Care
www.carboncalculator.org
Commercial company selling
carbon offsets, but very
useful online carbon
calculator

Carbon Neutral Company
www.carbonneutral.com
0207 833 6000
Another carbon offsets
company, again with useful
carbon calculator online

EU carbon calculator
www.mycarbonfootprint.eu
European site suggesting
carbon reduction tips for
households

**Low-impact Living
Initiative**
www.lowimpact.org
01296 714814
Everything from composting
to solar panels, with online
shop and focus on DIY

National Energy Foundation
www.nef.org.uk
01908 665555
Focusing on sustainable energy solutions for the UK

OneWorld Carbon Countdown
http://uk.oneworld.net/section/blogs/carbon
Carbon and climate 'mega-blog'

Friends of the Earth – Living
www.foe.co.uk/living/index.s html
FoE general number: 0207 490 1555
'Living green' section from the environmental campaign group

ACKNOWLEDGEMENTS

Many thanks to George Marshall of the Climate Outreach and Information Network for a thorough examination of an early draft of this book, and many helpful suggestions.

I am grateful too to Mayer Hillman, whose own work on carbon calculations and rationing has been invaluable.

Thank also to the Energy Saving Trust for their voluminous information, some of it reproduced here, and to the Centre for Alternative Technology for the same. Any errors or omissions, of course, remain the responsibility of the author.

FURTHER READING

- *Climate change begins at home: Life on the two-way street of global warming*, by Dave Reay, published by Macmillan, 2006 paperback
Climate change and its solutions through the eyes of a typical family

- *How we can save the planet*, by Mayer Hillman with Tina Fawcett, published by Penguin, paperback 2004
Why we need carbon rationing, and how to get your personal footprint down

- *The little green book of big ideas*, by Friends of the Earth and Think Publishing
Pocket-sized book with 185 eco-friendly tips and ideas

- *The energy saving house*, by Centre for Alternative Technology and Friends of the Earth, 2003
How to reduce your energy consumption at home

- *High Tide: How climate crisis is engulfing our planet*, by Mark Lynas, Harper Perennial, 2005 paperback
Still the best climate change book ever written – honest!

- *Heat: How to stop the planet burning*, by George Monbiot, Penguin, 2006
How we can cut carbon emissions by 90% by 2030 without causing an insurrection

- *The Last Generation: How nature will take her revenge for climate change*, by Fred Pearce, Eden project books, 2006 paperback
Why climate change is the most dangerous issue humanity has ever faced

- *Six Degrees*, by Mark Lynas, Fourth Estate, 2007
A degree by degree guide to global warming impacts over the coming century

Your carbon diary

	2007
Your total gas footprint	.... kilograms
Your total heating oil footprint	.... kilograms
Your total coal footprint	.... kilograms
Your total electrical carbon footprint	.... kilograms
Your total carbon footprint from driving	.... kilograms
Your total carbon footprint from motorbike riding	.... kilograms
Your total carbon emissions from public transport	.... kilograms
Your total carbon emissions from flights	.... kilograms
Your carbon footprint from food and other consumption	.... kilograms
Grand total for 2007	............kilograms

	2008
Your total gas footprint	.... kilograms
Your total heating oil footprint	.... kilograms
Your total coal footprint	.... kilograms
Your total electrical carbon footprint	.... kilograms
Your total carbon footprint from driving	.... kilograms
Your total carbon footprint from motorbike riding	.... kilograms
Your total carbon emissions from public transport	.... kilograms
Your total carbon emissions from flights	.... kilograms
Your carbon footprint from food and other consumption	.... kilograms
Grand total for 2008	............kilograms

2009

Your total gas footprint	.... kilograms
Your total heating oil footprint	.... kilograms
Your total coal footprint	.... kilograms
Your total electrical carbon footprint	.... kilograms
Your total carbon footprint from driving	.... kilograms
Your total carbon footprint from motorbike riding	.... kilograms
Your total carbon emissions from public transport	.... kilograms
Your total carbon emissions from flights	.... kilograms
Your carbon footprint from food and other consumption	.... kilograms
Grand total for 2009	...........kilograms

2010

Your total gas footprint	.... kilograms
Your total heating oil footprint	.... kilograms
Your total coal footprint	.... kilograms
Your total electrical carbon footprint	.... kilograms
Your total carbon footprint from driving	.... kilograms
Your total carbon footprint from motorbike riding	.... kilograms
Your total carbon emissions from public transport	.... kilograms
Your total carbon emissions from flights	.... kilograms
Your carbon footprint from food and other consumption	.... kilograms
Grand total for 2010	............kilograms

	2011
Your total gas footprint	.... kilograms
Your total heating oil footprint	.... kilograms
Your total coal footprint	.... kilograms
Your total electrical carbon footprint	.... kilograms
Your total carbon footprint from driving	.... kilograms
Your total carbon footprint from motorbike riding	.... kilograms
Your total carbon emissions from public transport	.... kilograms
Your total carbon emissions from flights	.... kilograms
Your carbon footprint from food and other consumption	.... kilograms
Grand total for 2011	............kilograms

2012

Your total gas footprint	.... kilograms
Your total heating oil footprint	.... kilograms
Your total coal footprint	.... kilograms
Your total electrical carbon footprint	.... kilograms
Your total carbon footprint from driving	.... kilograms
Your total carbon footprint from motorbike riding	.... kilograms
Your total carbon emissions from public transport	.... kilograms
Your total carbon emissions from flights	.... kilograms
Your carbon footprint from food and other consumption	.... kilograms
Grand total for 2012	............kilograms

	2013
Your total gas footprint	.... kilograms
Your total heating oil footprint	.... kilograms
Your total coal footprint	.... kilograms
Your total electrical carbon footprint	.... kilograms
Your total carbon footprint from driving	.... kilograms
Your total carbon footprint from motorbike riding	.... kilograms
Your total carbon emissions from public transport	.... kilograms
Your total carbon emissions from flights	.... kilograms
Your carbon footprint from food and other consumption	.... kilograms
Grand total for 2013	............kilograms

	2014
Your total gas footprint	.... kilograms
Your total heating oil footprint	.... kilograms
Your total coal footprint	.... kilograms
Your total electrical carbon footprint	.... kilograms
Your total carbon footprint from driving	.... kilograms
Your total carbon footprint from motorbike riding	.... kilograms
Your total carbon emissions from public transport	.... kilograms
Your total carbon emissions from flights	.... kilograms
Your carbon footprint from food and other consumption	.... kilograms
Grand total for 2014	............kilograms

2015	
Your total gas footprint	.... kilograms
Your total heating oil footprint	.... kilograms
Your total coal footprint	.... kilograms
Your total electrical carbon footprint	.... kilograms
Your total carbon footprint from driving	.... kilograms
Your total carbon footprint from motorbike riding	.... kilograms
Your total carbon emissions from public transport	.... kilograms
Your total carbon emissions from flights	.... kilograms
Your carbon footprint from food and other consumption	.... kilograms
Grand total for 2015	............kilograms

Useful websites

HOUSEHOLD EMISSIONS

www.secondnatureuk.com

www.excelfibre.com

www.est.org.uk/myhome/localadvice

www.heca.co.uk

www.est.org.uk/myhome/gid

www.boilers.co.uk

www.corgi-gas-safety.com

www.est.org.uk/myhome/efficientproducts/recommended

www.electrisave.co.uk

www.greenelectricity.org

www.oneclickpower.com

www.stopurban4x4s.org.uk

TRANSPORT EMISSIONS

www.vcacarfueldata.org.uk/search/search.asp

www.lcc.org.uk

http://critical-mass.info

www.saferoutestoschools.org.uk

www.ctc.org.uk

Useful websites

www.sustrans.org.uk

www.transportdirect.info

www.nationalrail.co.uk

http://chooseclimate.org/flying

www.carbonneutral.com/cncalculators/flightcalculator.asp

www.carboncalculator.org

http://coinet.org.uk/projects/challenge/measure

FOOD AND CONSUMPTION

www.farmersmarkets.net

www.whyorganic.org

www.nsalg.org.uk

www.cat.org.uk/catpubs/display.tmpl?sku=ts_cs

www.carbontrust.co.uk

www.lowcarbonbuildings.org.uk

SUSTAINABILITY

www.lowcarbonbuildings.org.uk/info/installers

www.solarsavers.co.uk

www.solarforlondon.org

Useful websites

www.lowimapact.org

www.good-energy.co.uk

www.lowcarbonbuildings.org.uk/myhome/generating/
types/solarpv

www.est.org.uk/schri

www.theyellowhouse.org.uk

www.bwea.com/noabl

www.windsave.com

www.stealthgen.com

www.windtrap.co.uk

www.renewabledevices.com

www.cat.org.uk

www.est.org.uk/myhome/generating/types/hydro

www.est.org.uk/myhome/generating/types/groundsource

www.gshp..org.uk

www.est.org.uk/myhome/generating/types/biomass

www.nef.org.uk/logpile/fuelsuppliers/index.htm

INDEX